ecommerce bestpractices

First Edition

ecommerce bestpractices
how to market, sell, and service customers with Internet technologies.

8230 Old Courthouse Rd. • Suite 500 • Vienna, VA 22182
tel 703.226.3800 • www.mcfadyen.com • info@mcfadyen.com

www.ecommerce-best-practices.com

eCommerce Best Practices
How to Market, Sell and Service Customers with Internet Technologies
First Edition
by McFadyen Solutions

Published by:

McFadyen Solutions
8230 Old Courthouse Rd.
Suite 500
Vienna, VA 22182
703.226.3800
info@mcfadyen.com
www.mcfadyen.com

Copyright © 2008 by McFadyen Solutions.
All rights reserved. No part of this book may be reproduced or transmitted in any form or by any means, electronic or mechanical, including photocopy, recording or by any information storage and retrieval system, without written permission from McFadyen Solutions, except for the inclusion of brief quotations in a review.

While the author/publisher has taken precautions in preparing this book, the author/publisher assumes no repsonsibility for errors or omissions, or for damages resulting from the use of the information contained herein.

McFadyen Solutions welcomes comments on this book. Please share your thoughts to info@mcfadyen.com.

All trademarks listed in this book are the property of their respective owners.

McFadyen Solutions books are available at special quantity discounts for corporate training, academic programs, or other special events.

Printed in the U.S.A.

ISBN 978-0-9815951-0-8

Library of Congress Control Number: 2008902140

eCommerce Best Practices

TABLE OF CONTENTS

PREFACE

INTRODUCTION

PART I: MARKETING
1.1 Search Engine Optimization
1.2 Email Marketing
1.3 Time-based Scenarios
1.4 Promotions
1.5 Rich Media
1.6 Branding
1.7 Social Networking
1.8 Affiliates
1.9 Sweepstakes
1.10 Newsletter

PART II: SELLING
2.1 Up-selling and Cross-selling
2.2 Shopping Cart
2.3 Checkout
2.4 Payment Processing
2.5 Fraud Prevention
2.6 Fulfillment / Shipping
2.7 Inventory Management
2.8 Merchandising
2.9 User Ratings / Reviews
2.10 Gift Services
2.11 B2B
2.12 Soft Goods
2.13 Gift Cards
2.14 Customer Loyalty Programs
2.15 Wish List
2.16 Order Management

PART III: SERVICE
3.1 Online Customer Support
3.2 Product Claims & Returns
3.3 Channel / Partner Communication
3.4 Monitored Forums
3.5 Warranty and Maintenance Contracts
3.6 Online User Documentation & Manuals
3.7 Customer Call Center Support & Service

PART IV: TECHNOLOGIES
4.1 UI/ UE
4.2 Integration
4.3 SOA
4.4 Information Architecture
4.5 eCommerce Platforms
4.6 Implementation Service Vendors
4.7 Infrastructure Security
4.8 Application Security
4.9 Infrastructure
4.10 Analytics/Business Intelligence
4.11 Personalization
4.12 Search
4.13 Chat
4.14 Blogs
4.15 RSS / PODCASTS
4.16 Portals

PART V: HOW TO
5.1 RFP
5.2 Implementation
5.3 Operations / Maintenance
5.4 Data Migration
5.5 Content Management
5.6 Holiday Season
5.7 Multi-Channel
5.8 ROI Analysis / Cost Benefits
5.9 Rough Order Of Magnitude Estimation
5.10 PCI Compliance
5.11 SaaS vs. Managed Service

eCommerce Best Practices

PREFACE		**XI**
INTRODUCTION		**XII**
MARKETING		**1**
1.1.1	**Search Engine Optimization & Marketing**	3
1.1.1	Introduction	3
1.1.2	Results Pages	3
1.1.3	Search Engine Optimization	4
1.1.4	Search Engine Marketing	4
1.1.5	Proven Techniques	5
1.1.6	Dangerous Techniques	6
1.1.7	Penalties	6
1.1.8	Measuring Success	7
1.2	**Email Marketing**	8
1.2.1	Introduction	8
1.2.2	Email Success Trio	8
1.2.3	Automating Your Emails	10
1.2.4	Customer Optimization for Email Marketing	12
1.2.5	What to Avoid	13
1.3	**Time-based Scenarios**	15
1.3.1	Introduction	15
1.3.2	Marketing and Scenarios	15
1.3.3	Advantages of Scenarios	16
1.3.4	What to Look for in Scenario Technology	17
1.4	**Promotions**	19
1.4.1	Introduction	19
1.4.2	Strategies that Make a Difference in Promotions	19
1.4.3	Some Quick Ideas for Increased Revenues	20
1.4.4	Some Low Cost Promotional Ideas	21
1.4.5	Promotional Types	22
1.5	**Rich Media**	24
1.5.1	Introduction	24
1.5.2	Does Your Site Need RIA?	24
1.5.3	Opportunities with RIA	25
1.5.4	Advantages of RIA	27
1.5.5	Disadvantages of RIA	28
1.5.6	Choosing an RIA Solution	29
1.6	**Branding**	30
1.6.1	Introduction	30
1.6.2	Branding for Internet Retailers	31
1.6.3	Dimensions of Branding—Offline vs. Online	33
1.6.4	Brand Personality	34
1.6.5	Online Commercials	35
1.6.6	Reach Out to Your Customers	36

eCommerce Best Practices

1.7	**Social Networking**	**38**
1.7.1	Introduction	38
1.7.2	Popular Mediums	39
1.7.3	What to Consider When Interacting with a Community	40
1.7.4	How to Convert Browsers to Buyers in Social Network	41
1.7.5	Analysis	42
1.8	**Affiliates**	**43**
1.8.1	Introduction	43
1.8.2	Benefits to a Retailer	44
1.8.3	Content is Key to Affiliate Marketing	44
1.8.4	What to Consider	45
1.8.5	How to Make Affiliate Marketing Work	46
1.8.6	Affiliate Tools—Reviews, RSS, Podcasts	47
1.9	**Sweepstakes**	**50**
1.9.1	Introduction	50
1.9.2	Sweepstakes Rules	51
1.9.3	Post-Sweepstakes Analysis	53
1.10	**Newsletter**	**54**
1.10.1	Introduction	54
1.10.2	Does Your Company Need Newsletters?	54
1.10.3	What to Include in the Newsletter	55
1.10.4	What to Measure from Newsletters	57

SELLING 60

2.1	**Up-selling and Cross-selling**	**61**
2.1.1	Introduction	61
2.1.2	What is Up-selling?	61
2.1.3	Up-selling Hints	62
2.1.4	What is Cross-selling?	63
2.1.5	Placement for Effective Selling	64
2.1.6	Up-sell and Cross-sell Success—Know Your Customer	65
2.2	**Shopping Cart**	**66**
2.2.1	Introduction	66
2.2.2	Shopping Cart Abandonment	66
2.2.3	Shopping Cart Functionalities	68
2.2.4	Multiple Shopping Carts	69
2.2.5	The Ideal Shopping Cart	70
2.3	**Checkout**	**72**
2.3.1	Introduction	72
2.3.2	Checkout Flow	72
2.3.3	Checkout Points to Consider	74
2.4	**Payment Processing**	**77**
2.4.1	Introduction	77
2.4.2	Terms and Definitions	78
2.4.3	Forms of Payment	80
2.4.4	Miscellaneous Methods and the Future	85
2.5	**Fraud Prevention**	**87**

2.5.1	Introduction	87
2.5.2	Fraud Prevention Techniques	88
2.5.3	Fraud Prevention Framework	92
2.6	**Fulfillment / Shipping**	**96**
2.6.1	Introduction	96
2.6.2	Importance and Components of Fulfillment	100
2.6.3	Fulfillment Process in Detail	102
2.6.4	Challenges and Suggestions	103
2.6.5	Future	105
2.7	**Inventory Management**	**106**
2.7.1	Introduction	106
2.7.2	Balancing Overstock and Understock Items	106
2.7.3	Suggested Methods	107
2.8	**Merchandising**	**111**
2.8.1	Introduction	111
2.8.2	Effective Techniques Used	111
2.8.3	Recommended Methods	112
2.9	**User Ratings / Reviews**	**115**
2.9.1	Introduction	115
2.9.2	Products in Review	116
2.9.3	Maintenance	117
2.9.4	Balance of Positive and Negative Reviews	117
2.10.1	**Gift Services**	**119**
2.10.2	Introduction	119
2.10.3	Some Common Gift Services	120
2.10.4	Customer Relationships with Gift Services	122
2.11.1	**B2B**	**123**
2.11.2	Introduction	123
2.11.3	B2B Attributes	124
2.11.4	Usage	124
2.11.5	Functionalities	125
2.11.6	Technology, Standards, Considerations	126
2.11.7	Functions and Tools for Support	126
2.11.8	B2B Exchange	127
2.11.9	Collaborative B2B eCommerce	128
2.11.10	B2B Benefits	128
2.11.11	Recommendations	129
2.12	**Soft Goods**	**130**
2.12.1	Introduction	130
2.12.2	Differences between Soft and Hard Goods (Consumer's POV)	130
2.12.3	Differences Between Soft and Hard Goods (Provider's POV)	132
2.12.4	Verticals that Deal Mostly in Soft Goods	132
2.12.5	Dealing with Piracy and Unauthorized Use of Soft Goods	132
2.13	**Gift Cards**	**135**
2.13.1	Introduction	135
2.13.2	What are Gift Cards?	135
2.13.3	Gift Card Benefits	136
2.13.4	Customized Design and Online Gift Cards	137

eCommerce Best Practices

	2.13.5	Terms and Conditions	138
2.14		**Customer Loyalty Programs**	**140**
	2.14.1	Introduction	140
	2.14.2	Common Strategies for Loyalty Programs	141
	2.14.3	Marketing Strategies	141
	2.14.4	Balance of Rewards and Loyalty	143
	2.14.5	Measure Customer Loyalty Required	143
	2.14.6	Identifying and What to Collect	144
2.15		**Wish Lists**	**145**
	2.15.1	Introduction	145
	2.15.2	What is a Wish List?	145
	2.15.3	Wish Lists Benefits	146
	2.15.4	Wish Lists vs. Gift Registries	147
2.16		**Order Management**	**149**
	2.16.1	Introduction	149
	2.16.2	Value of Order Management	149
	2.16.3	Multi-Channel and Inventory Control for Order Management	151
	2.16.4	Steps to Selecting an Order Management System	152

SERVICE 156

3.1		**Online Customer Support**	**157**
	3.1.1	Introduction	157
	3.1.2	Attributes and Features	157
	3.1.3	Technologies and Issues	159
	3.1.4	Recommendations	160
3.2		**Product Claims & Returns**	**161**
	3.2.1	Introduction	161
	3.2.2	Reasons for Returns	161
	3.2.3	Solutions Provided by Software	162
	3.2.4	How to Reduce Returns	163
3.3		**Channel / Partner Communication**	**166**
	3.3.1	Introduction	166
	3.3.2	Extranet and Its Benefits	166
	3.3.3	Channel and Partnerships	167
3.4		**Monitored Forums**	**169**
	3.4.1	Introduction	169
	3.4.2	Pros and Cons	169
	3.4.3	Desired Features	170
	3.4.4	Third Party Tools	172
3.5		**Warranty and Maintenance Contracts**	**173**
	3.5.1	Introduction	173
	3.5.2	Warranties	173
	3.5.3	Maintenance Contracts	174
	3.5.4	Maintenance Checklist	174
3.6		**Online User Documentation & Manuals**	**176**
	3.6.1	Introduction	176

3.6.2	Need for Online Manuals	176
3.6.3	Content Distribution	177
3.7	**Customer Call Center Support & Service**	**178**
3.7.1	Introduction	178
3.7.2	Investment, not Expense	178
3.7.3	CRM and CSR Training	179
3.7.4	Click-to-Call / Click-to-Chat	180
3.7.5	Maintaining Customer Service	181
3.7.6	Call Center Optimization	181

TECHNOLOGIES 184

4.1	**User Interface / User Experience**	**185**
4.1.1	Introduction	185
4.1.2	Considerations	185
4.1.3	Best Practices	186
4.2	**Integration**	**190**
4.2.1	Introduction	190
4.2.2	Data Integration	191
4.2.3	Web Services, SOAP, WSDL, and UDDI	191
4.2.4	Services-Oriented Architecture (SOA)	193
4.2.5	Enterprise Service Bus	194
4.3	**Service Oriented Architecture (SOA)**	**195**
4.3.1	Introduction	195
4.3.2	SOA Key End Points	196
4.3.3	Suitability	197
4.3.4	Guidelines	197
4.3.5	Benefits of SOA	198
4.3.6	SOA and Web Services	199
4.3.7	eCommerce and SOA	200
4.4	**Application Architecture**	**201**
4.4.1	Introduction	201
4.4.2	One-Tier Architecture	201
4.4.3	Two-Tier Architecture	201
4.4.4	Three-Tier Architecture	202
4.4.5	N-Tier Architecture	203
4.4.6	MVC Architecture	203
4.5	**eCommerce Platforms**	**206**
4.5.1	Introduction	206
4.5.2	Perpetual License Vendors	206
4.5.3	Considerations before Selecting Service Vendors	207
4.5.4	Selecting the Right eCommerce System	208
4.5.5	When Product and Service Vendors Collide	209
4.5.6	Pay Pal, Google Checkout and Small Businesses	210
4.6	**Implementation Service Vendors**	**212**
4.6.1	Introduction	212
4.6.2	Advantages and Disadvantages to Service Providers	213
4.6.3	Offshore IT Outsourcing	214

eCommerce Best Practices

4.6.4	Managed Services	214
4.7	**Infrastructure Security**	**215**
4.7.1	Introduction	215
4.7.2	Credit Card Security	215
4.7.3	Traditional Mechanisms for Network Security & Their Drawbacks	216
4.7.4	Vulnerabilities Associated with Database Security	217
4.8	**Application Security**	**219**
4.8.1	Introduction	219
4.8.2	OWASP	219
4.8.3	SSL & HTTPS	221
4.9	**Infrastructure**	**222**
4.9.1	Introduction	222
4.9.2	Infrastructural Components	223
4.9.3	Hosting	225
4.9.4	Applications	226
4.9.5	Platforms	226
4.9.6	Monitoring and Support	227
4.10	**Analytics / Business Intelligence**	**229**
4.10.1	Introduction	229
4.10.2	Web Analytics	230
4.10.3	Data Analytics	230
4.10.4	Conversion Funnel	231
4.10.5	Actual Behavior (A/B) Testing	231
4.10.6	Key Performance Indicators	232
4.10.7	Tools Category	232
4.11	**Personalization**	**234**
4.11.1	Introduction	234
4.11.2	Why Personalize?	234
4.11.3	Elements of Personalization	235
4.11.4	Building User Profiles	236
4.11.5	Content Delivery	237
4.11.6	Tips for Personalization	238
4.11.7	Personalization — A Myth?	239
4.12	**Search**	**240**
4.12.1	Introduction	240
4.12.2	Faceted Search	240
4.12.3	Consumer Insight	241
4.12.4	Discovering New Items	242
4.12.5	Search UI Considerations	243
4.13	**Chat**	**244**
4.13.1	Introduction	244
4.13.2	Desired Functionality	244
4.13.3	Pros and Cons	245
4.13.4	Desired Features	246
4.13.5	Third Party Tools	248
4.13.6	Curreunt State	248
4.14	**Blogs**	**250**
4.14.1	Introduction	250

4.14.2	Fostering Customer Relationships	250
4.14.3	Reasons to Blog	251
4.14.4	Blog Netiquette	252
4.14.5	Who Runs Your Blog?	253
4.14.6	Getting Started	253
4.14.7	Monitoring Risk	254
4.14.8	Feedback	255

4.15 RSS / PODCASTS 256
- 4.15.1 Introduction 256
- 4.15.2 RSS vs. Email 256
- 4.15.3 RSS Best Practices for Feed Producers 257
- 4.15.4 RSS Best Practices for Feed Consumers 257
- 4.15.5 Podcasting 257
- 4.15.6 RSS and eCommerce 258

4.16 Portals 260
- 4.16.1 Introduction 260
- 4.16.2 Modernize Your Business 261
- 4.16.3 Benefits of Deploying Portals 262
- 4.16.4 Portals and Their Impact on an Organization 263
- 4.16.5 Layout of Typical Portal Implementation 264
- 4.16.6 Shared Portlets and Community Portlets 266

HOW-TO 268

5.1 RFP 269
- 5.1.1 Introduction 269
- 5.1.2 Checklist in the RFP Process 269
- 5.1.3 Flaws in the RFP Process 271
- 5.1.4 Sample RFP 272
- 5.1.5 Limitations to RFPs 273
- 5.1.6 RFI as a Solution 274

5.2 Implementation 275
- 5.2.1 Introduction 275
- 5.2.2 Design Patterns 275
- 5.2.3 Reusing Code 276
- 5.2.4 Automation Tools 277

5.3 Operations / Maintenance 279
- 5.3.1 Introduction 279
- 5.3.2 Maintenance 281
- 5.3.3 Operations in Returns 282
- 5.3.4 Operations in Accounting 283
- 5.3.5 Operations in Marketing 284

5.4 Data Migration 285
- 5.4.1 Introduction 285
- 5.4.2 Tools for Data Migration 288
- 5.4.3 Issues with Data Migration 289
- 5.4.5 Planning 290

5.5 Content Management 293

eCommerce Best Practices

5.5.1	Introduction	293
5.5.2	Do You Need CMS?	293
5.5.3	How to Find the Right Vendor	295
5.5.4	Considerations	296
5.5.5	Lessons Learned	299
5.6	**Holiday Season**	**303**
5.6.1	Introduction	303
5.6.2	What to Implement for the Holiday Season	303
5.6.3	Testing	307
5.6.4	Marketing and IT Synchronization	309
5.7	**Multi-Channels**	**310**
5.7.1	Introduction	310
5.7.2	Operations in Multi-Channels	311
5.7.3	Consistency in Multi-Channels	312
5.7.4	Core Competency	312
5.7.5	Customer Service in Multi-channels	314
5.8	**ROI Analysis / Cost Analysis**	**316**
5.8.1	Introduction	316
5.8.2	Organizational ROI Mindset	316
5.8.3	ROI in Analytic Tools	318
5.8.4	ROI in Behavioral Targeting	319
5.9	**Rough Order Of Magnitude (ROM) Estimation**	**320**
5.9.1	Introduction	320
5.9.2	What is Rough Order of Magnitude (ROM)?	321
5.9.3	Whose Fault is it anyway?	321
5.9.4	A Methodology for Estimation	322
5.9.5	Analysis of the Methodology	324
5.9.6	Best & Worst Practices	325
5.9.7	How to Use a ROM Estimate	325
5.10	**PCI Compliance**	**326**
5.10.1	Introduction	326
5.10.2	Risks	326
5.10.3	Requirements	327
5.10.4	Evaluation Processes	329
5.10.5	Validation	330
5.11	**SaaS vs. Managed Service**	**332**
5.11.1	Introduction	332
5.11.2	SaaS (OnDemand)	333
5.11.3	Managed Service Providers and Perpetual Licensing	334
5.11.4	Budgeting for SaaS and Perpetual Licensing	335
GLOSSARY		**338**
INDEX		**357**

PREFACE

By Tom McFadyen, President of McFadyen Solutions

Enterprise-grade eCommerce covers a broad and diverse range of topics. This book addresses as many of those topics as possible in a few hundred pages. Many of the chapters could be expanded into a book by themselves. The goal of **eCommerce Best Practices** is to provide a high-level overview across a number of eCommerce areas.

I'd like to personally thank the many hard-working employees of McFadyen Solutions who have developed many of these best practices and our valuable customers and partners with whom they've been implemented, tested and refined. This book and McFadyen Solutions would not exist without them. For over a decade McFadyen Solutions has been focused on implementing large-scale eCommerce solutions. Our global team has implemented some of the largest eCommerce sites in the world - many with over 1,000,000 SKU's, 1,000,000 users, 1,000,000 dollars/day sales, 1,000,000 order-items shipped per day, and 1,000,000 web assets under management.

eCommerce is a relatively young and evolving industry compared to general commerce which has been evolving for thousands of years. Look for updates to this first edition book as the industry continues to evolve and new best practices are formed. Also check www.eCommerce-Best-Practices.com for updates as future editions of the book are written.

We also welcome your feedback on the content in this book. Please contact us at info@mcfadyen.com or call (703)226-3800 with your comments.

INTRODUCTION

Anybody can setup a shopping cart online. However the solution needed for real-world business quickly gets far more complex.

For example, say a customer buys four items on your site. They use a combination of a gift card (stored value card) and a credit card for payment. Two items are shipped to one address and two go to another address. One product comes from a wedding registry that must be updated and also needs to be gift wrapped. Two of the items are available in one fulfillment center, one in another center, and the fourth is on back-order. A 20% discount applies because of the total dollar volume of the purchase, but one item is later returned to a store reducing the total order size below the discount threshold. The customer then calls to find other products to match their prior purchases and expects the call center agent to know their purchase history and offer compatible cross-sells.

These are just a few of the topics which must be mastered for successful enterprise eCommerce. Obviously eCommerce is not a single feature (e.g. shopping cart). It is a customer-engagement lifecycle (**market, sell, service**) that must be supported by processes (**how-to**) and tools (**technologies**). This book is divided into five parts to cover these topics:

Part I - Marketing an eCommerce site sets the stage for your company and brand to position and differentiate itself from the competition. This includes Search Engine Optimization, often the biggest topic for Marketing teams, but also includes email marketing, social networks, and rich media among others. These resources and actions will help you maximize your branding efforts and drive highly-qualified traffic to your site.

Part II - Selling discusses how to maximize conversion rates and order size while reducing fraud and fulfillment costs. Some features highlighted in this section are up-selling/cross-selling, user ratings/reviews, merchandising, gift cards, and customer loyalty programs.

Part III - Service continues the customer relationship after the sale. Not only does effective online customer support reduce costs, but it also provides great opportunities for additional cross-selling. Topics included in this section include developing a versatile customer relationship management system,

eCommerce Best Practices

offering a comprehensive knowledgebase, and providing a forum for open discussions.

Part IV - Technology describes the best solutions that will make your site a reality and keep it operating and scaling appropriately. Architecture, SOA and Security are some of the hot-topics for the online market. This section will also touch on the newest technologies that can enhance your site within the market. These technologies include blogs, RSS and podcasts, as well as faceted site search.

Part V - How-To will give you more direction in the process of developing and operating your eCommerce system. This section covers the best practices on Data Migration, Multi-Channel integration, and Rough Order of Magnitude Estimation.

eCommerce Best Practices

I

marketing

ecommerce
bestpractices

MARKETING

There are a number of direct correlations between online marketing techniques and direct customer facing sales tactics. In both cases, critical components of success include fundamental exercises like managing product exposure, building brand awareness and establishing a firm position in their target marketplace. Despite these similarities, marketing an organization's product and services online requires a slightly more customized approach due to the demand of today's tech-savvy customer.

In this section, you will learn about some of the following techniques and suggested best practice approaches to implementing them:

- Importance of Search Engine Optimization
- Advantages of Email Marketing
- Benefits of Time-based Scenarios
- Implementing Promotions
- How Rich Media affects the Online Community
- Why Branding is Important for Growth
- Social Networks and their Influence
- What to Consider when Creating Affiliations
- Impact of Sweepstakes on the Shopper
- Importance of Newsletters

eCommerce Best Practices

1.1.1 Search Engine Optimization & Marketing

1.1.1	Introduction
1.1.2	Result Pages
1.1.3	Search Engine Optimization
1.1.4	Search Engine Marketing
1.1.5	Proven Techniques
1.1.6	Dangerous Techniques
1.1.7	Penalties
1.1.8	Measuring Success

1.1.1 Introduction

One of the most effective and powerful strategies available today for any website including retail websites is expanding your online business through search engine placement. Since their arrival in the early '90s, organizations have been struggling to promote their online projects on search engine results pages. Even today, there is still confusion amongst retailers regarding the best methods and priorities for establishing and maintaining an optimal search engine presence.

Each of the "Big Three" players in the search engine market – Google, Yahoo and Microsoft – constantly tweak and refine their product to meet and anticipate the rapidly evolving needs and desires of web searchers. Anyone with a website wants to show up as the first, most prominent link to an infinite variety of search queries, but only a handful of links make it on the cherished first page.

1.1.2 Results Pages

Whenever someone enters a word or phrase into a search engine, they generate a list of results, sorted by relevance. In most cases, it also shows a small collection of advertisements specifically written for that exact search query. But how does the search engine decide which pages and advertisements are the most important—the most relevant?

Search engines use a dizzying array of factors to compile that list of results, addressing with almost every aspect of the page itself, and its relationship to many other web pages that point to it. The popularity and relevance of pages

eCommerce Best Practices

are judged based on clearly defined, but largely secret algorithms. Each results page is also host to a group of mini-commercials, which are also selected by relevance, but are additionally weighted by the amount of money their publisher is willing to pay for placement.

1.1.3 Search Engine Optimization

Search Engine Optimization is the pursuit of the perfectly constructed, machine readable web page that accurately and effectively reflects the content and goals of that particular document to the various search engines. The desired result of effective SEO is appearing in a prominent location in search engine results pages when the appropriate words or phrases are entered. Initial success is measured in positional improvements in the page's standard – or organic – search results, but the ultimate goal is to drive visitors – or clicks – to the page in question.

Internal SEO Factors

The best SEO results are achieved through the combination of a wide variety of disciplines that come together to craft a "search optimized" page: descriptive writing, precise HTML coding, and intimate knowledge of your target audience. A failure in any of these three areas could seriously weaken any SEO initiative.

External SEO Factors

A fourth factor that is much more difficult to influence is each search engine's opinion of the relative worth or popularity of a page. Each search engine's formula is unique – and secret – but they all factor in the number of links to a page, and the relative worth of those linking pages. Other secondary indicators may include the frequency of page updates, the rate at which links to the page are added, and even the age of the domain name.

1.1.4 Search Engine Marketing

Sometimes, having a well formed page isn't enough to quickly secure a top spot in the SERPs, or perhaps a page just can't rank well for a desired group of words or phrases. The ability to craft a short snippet of text that begs to be clicked, coupled with the paid inclusion programs of all the major search engines offers a powerful alternative.

Paid search results usually appear above or beside the search engine's more numerous organic search results, with the highest bidder for a particular keyword or keyphrase right on top. While some users may be reluctant to click a paid search ad, many more will follow paid links if they appear to be relevant to the provided query.

1.1.5 Proven Techniques

Every search engine has a ranking algorithm and all the search spiders are tuned to the same. Web spiders are on a mission to identify quality content and separate it from irrelevant information, grab the content, extract the essence of this content on a page-by-page basis, compare content for source reputation, determine the relationship between the content and site and finally assign the result. The algorithm uses rules to identify the most relevant pages depending on your page text content and its context. The context is usually indicated by the number of links from other pages and sites. To deliver your site to a search engine result, there are some common factors needed.

> **Though paid search results are less popular, users are more likely to click on them if they are relevant to their query.**

Page Optimization
This is where you match copy on the page to the search items entered. Factors are dependent upon the keywords used as anchor text, page title tags, meta-tags, etc. Make sure that you do not overdo the keywords matching copy on your site as otherwise it would look like spamming. It is more like labeling the content on your site with the headline and link text reflecting the editorial content.

Link Building
Your website ranking depends on inbound and outbound links. Search spiders access each link to a page from another page within your site and other sites as well. Pages that contain more inbound links are usually ranked high, but again you need to be careful that these outbound links are relevant to the content of your site. Essentially, the quality of these links becomes a governing factor.

Content is the Key
Your site requires content that strikes a balance between information on the pages which the user reads and keywords which the search spiders effectively translate to position you at the top result page. Content is vital - and it should

come from within key departments in your company like Media relations or PR or Marketing.

1.1.6 Dangerous Techniques

Some SEO practices intended to enhance a website's position in the SERPs are explicitly forbidden by various search engines. Others are merely ill-advised.

Cloaking: This is the technique where web sites intercept each request, attempt to distinguish the ones from search engine bots, and display content significantly different that that seen by the end user. It is obvious how this can be misused; you can load the web pages you serve up to the bots with misinformation to get a higher ranking. If detected, the offending domain could be pushed farther down the list of results, or removed entirely.

Link Farms: Being a part of link farms, where web masters agree to put links to each other's sites on their pages to increase their ranking, is also a quick way to get blacklisted.

Hidden keywords: A primitive technique of pasting huge lists of desired keywords in very small type or otherwise invisible areas of a web page. Search engines have long-ago learned to detect most versions of this early SEO trick, yet it still occasionally shows up.

1.1.7 Penalties

Search engines are constantly updating their algorithms to improve the quality of their search results, and this includes punishing websites that seem to be cheating the system. It's not always easy to determine if a website has been penalized, as search engines often change the formula they use to evaluate and rank web pages. So a sudden drop in the SERPs is just as likely to be the result of an algorithmic change than a true penalty.

Currently, the only search engine that will confess to penalizing a particular domain is Google, which includes this information to verified webmasters as a part of its Webmaster Tools utility, available at http://www.google.com/webmasters/tools .

Before beginning an optimization project, carefully review the guidelines of the major search engines. Their rules may seem arbitrary, and they are not always enforced, but websites ignore these rules at their peril.

1.1.8 Measuring Success

Due to the secrecy surrounding the various search engine algorythms, Search Engine Optimization and Marketing are inexact sciences. Even SEM successes, which are driven heavily by bid price, are also sensitive to proper keyword targeting and effective copywriting. Determining the relative worth of the various SEO techniques is part science, and part art.

Ultimately, the true measure of success is based on the goals established at the beginning of the project. It should come as no surprise that increased sales are quote often the primary goal of any eCommerce optimization and marketing campaign, but improved brand awareness and increased visits from a target demographic are also important achievements.

Whatever your goals or budgetary limits, search engine optimization and marketing are critical considerations for the launch and continued success of any web-based endevor.

eCommerce Best Practices

1.2 Email Marketing

1.2.1   Introduction
1.2.2   Email Success Trio
1.2.3   Automating Your Emails
1.2.4   Customer Optimization for Email Marketing
1.2.5   What to Avoid

1.2.1 Introduction

In terms of merchandising, email messaging and marketing is a powerful tool for measuring customer response to your site. As the online shopper becomes savvier, they will be able to quickly delineate which of their emails from specific retailers most interests them. Their inbox is increasingly becoming more saturated with emails with a variety of messages. Some promote the latest sale, new arrivals, or special offers for select shoppers. Some of these messages may be effective and result in them clicking on the link to the site, but an increasing number are not. A combination of relevancy and timing is important to the success of any email marketing campaign. So how do you ensure that your email marketing messages live up to the expectations?

1.2.2 Email Success Trio

Customer personalization is mandatory. Frequency should be regulated. Relevancy is critical. All of these three factors are extremely important to the overall success email marketing campaigns and should be ironed out before your email campaign is launched.

Personalization

Just because an email message includes the customer's name, do not think that it would constitute enough personalization to make it effective. Personalization starts with user information. This generally requires the user to register to the site in some capacity—name, email address, gender, etc. as opposed to being an anonymous user. As the customer shops, his or her user profile is being simultaneously generated. By measuring the customer's shopping behavior,

> **Personalization gives recipients a more customized experience than a general mass email would, thus increasing conversion rates.**

Part I: Marketing

eCommerce Best Practices

you can determine his or her interests to create a more personalized message. The objective is to obtain the most information possible about the consumer because it adds value to the primary message you want to communicate; be it a sale or a promotional offer.

Of course you must be wondering how you are going to customize each email message to every customer. It is possible with the right set of tools—having a dynamic content engine and customer database. With a dynamic content engine, marketers can create a single email template and populate it with specific customer information to deliver personalized messages. Your dynamic content engine requires a well segmented customer database and this data is collected over time with online surveys and customer purchase history opt-in forms.

Frequency

How often should you send email messages to your consumers? Every hour? Every day? Once a week? Many retailers worry whether information sent too frequently will lose its value over time because of its predictability, or worse, completely alienate customers. However, if the frequency of the emails is spaced out over a longer period of time, is that lack of brand presence hurting potential sales? Fundamentally, it is all about finding, and creating, the sweet spot—the perfect combination of timing, relevancy, and desire in your customers. Frequency can influence the desire of products. The more a customer sees your brand and the products you offer, the more top of mind awareness he or she has for making a purchase. It is important to test out different schedules of email deployment, while tracking the reactions and behaviors of the customers you send your campaigns to. You may want to divide your customers into specific user categories, which are then used to determine the optimal frequency of emails for that category. There is no such thing as over testing—as it is the key to a successful and profitable eCommerce site.

Relevance

No matter how personalized or frequent the emails are, none of that will matter unless the information is relevant and customized to the person receiving that message. If a customer has been searching for winter coats and they receive emails about watches, undoubtedly you may have just lost a sales opportunity, and, in fact, he or she might not

> **Email messages are only effective if they are relevant. User profiles and behavior monitoring will give you an understanding of each customer's interests.**

eCommerce Best Practices

even bother to open up the email. Be as focused as possible in your messaging. The information you gathered from each user will lend itself to forming a specific message that will peak their interest. Generalized promotions or sales may not result in higher conversion rates as a targeted message would. Sometimes customers are lack a clear shopping objective, and yet when they see an email message they suddenly become aware of how it could be relevant to them. This is due to personalization and being able to tap into your consumers' shopping behavior and purchasing/searching patterns. Therefore, every message should have meaning, and more so, should mean something to that consumer.

Each of these three factors is dependent on one another; almost cyclical in a sense. Personalization is the process of turning knowledge gained about your consumers into a marketing strategy that targets their needs, wants and behavior. Relevancy comes out of personalization because the marketer knows what is important to the consumer and can therefore make suggestions of products that the consumer might want or need. Frequency stems from relevancy because the marketer should not only know what is important to the consumer, but should also know when the consumer is most vulnerable to the messages.

With the above trio as the main driving force, you should also look at other options to elevate your email marketing success. For example acquiring more email addresses as part of your list building process. To do this you need quick and easy forms where a user does not need to spend a lot of time on registration. Strategically positioning these easy to use subscription slots are also important especially on high traffic pages on your site.

> **Make the registration process to your site as painless as possible. Only ask for essential information like Name and email address.**

1.2.3 Automating Your Emails

Automated emails are equally successful, especially when a customer's interest is high. These may include, but are not limited to welcome emails, shopping cart rescues, birthday and anniversary offers, and win-back offers. So how do you segment the types of emails based on your customer's age, interests, and interaction frequency? One option is to send emails not when you have promotions or campaigns but when the customer has performed a certain action. For example:

Welcoming Messages

Whenever your consumer uses the opt-in subscription or registration process, automated welcome messages add a personal touch. A little hospitality can go a long way.

Trigger Email Messages

Though initiated by a customer action or event, these messages are expected and communicated through a planned marketing opportunity. If a customer fails to complete their shopping process or does not remember that he or she has a 'buy 1 and get___ free,' *BOGO* offer, these automated trigger emails are useful.

Post-Purchase Emails

After customers have finished a shopping process, email messages related to their transaction details, shipping status, or delivery status can be automated and provided at frequent intervals to assure the customer that their purchase is on its way.

Birthday/Anniversary Email Messages

Still a very effective email marketing strategy, you can always target your customers during special occasions. Email greetings that offer customers' a gift voucher, dollar-off voucher, or a discount are effective for increasing your conversion rate. This, of course, is contingent upon the degree of personal information and significant dates the consumer is willing to offer up.

Viral Email Messages

Forwarding or referring email messages to friends is an extremely effective viral marketing tool. Your reach is extended, which results in more prospective consumers and exposure to a wider audience.

Wish List Emails

Most customers use their wish list basket to track products they want to purchase later—like a product bookmark. You could send email messages that are reminders or updates to what is in the wish list, which will encourage customers to purchase the products.

Customer Service Thank-You

Always ensure that whenever your customer has a dialogue with one of your customer service representatives, to send a thank-you email. Communicating that the time they spend with them had not gone unnoticed is affective in encouraging repeat visits.

Choice Methods

If your customer chooses to unsubscribe method for your newsletter, provide them with alternatives like RSS or offline methods. It is better to give choices to them than no choice at all; after all, your customers might change their mind and go with an alternative method of receiving information.

Creating these types of scenarios that require customer action is a powerful tool that enables the customer to ostensibly decide what messages he or she would be given.

1.2.4 Customer Optimization for Email Marketing

How do you ensure that the vast email database you have, can create internal marketing intelligence, build strategies, and communicate with customers more effectively? If you follow the methodology described below, you can optimize revenues per consumer and increase sales.

Always review and analyze your current customer base and the prospect database. Your customer database should be centralized so that you and the marketing team get a complete and objective view of the customer, their profile details, and behavioral data. This helps you to create and deliver tailored, relevant messaging campaigns.

Segmentation is very important. Based on each customer's purchase cycle and shopping patterns, segment your consumers based on models and individual attributes.

> **Segmentation of users for different email campaigns will help you get more targeted messages out to relevant customers.**

Because customers are becoming more aware and cognizant of online shopping "traps" in terms of their shopping behavior, you should implement analytic tools that will help you deconstruct customer intelligence and, thus, create more effective messages when sending your email campaigns.

Map out a strategy based on those who clicked-thru to the emails versus those who did not click-thru. You could test additional variables or conduct surveys to find out what was missing for the people who did not respond. Provide additional discounts if they have already purchased something or provide incentives and bonus points if they intend to make a purchase within a stipulated time. To get an effective feed on these strategies, it is suggested that you also set up system triggers and automate your messaging to each and every strategy map.

Every marketing medium within your strategy should integrate with other media, which is the multi-channel approach. Your customers can book and pay for products online but they also have the option of picking them up at physical retail locations.

Service notifications are important to maintain a dialogue with the customers. They should receive "Thank You" or confirmation messages, cross-sell and up-sell messages, shipping notices and delivery messages. Customer expectations and perceptions of your brand are positively enhanced when these types of notifications are sent.

1.2.5   What to Avoid

There is limited argument that emails play a significant role in driving online sales and website traffic. You always have to ask questions—how do you get the user to open your email, read them, respond and, ultimately, how do you increase conversion rates from email marketing? To answer this we must look at which aspects of the email practice could bring about negative responses.

Spamming

Sending out email messages to millions of addresses is by no definition a standard email marketing strategy. This is called SPAM and is recommended not to be used under any circumstances. SPAM results end up clogging the servers and bandwidth and are punishable by law. You take an unnecessary risk if you engage in Spamming.

Copy and Layout

Careful phrasing and design should always be a major factor in developing your email. You want it to be engaging and demand attention. If a customer

has to spend more than a couple of minutes to decipher the message, than clearly you must have done something wrong. Make sure your messages are communicated clearly and attract customers' attention.

eCommerce Best Practices

1.3 Time-based Scenarios

1.3.1 Introduction
1.3.2 Marketing and Scenarios
1.3.3 Advantages of Scenarios
1.3.4 What to Look for in Scenario Technology

1.3.1 Introduction

Scenarios, as mentioned briefly in the last chapter, are powerful tools for eCommerce retailers because it enables them to quickly adapt marketing messages to customer behavior. Scenarios are pre-defined series of interactions that are developed by various retailers, but are ultimately driven by the customers themselves. Every action a customer performs in regards to the eCommerce site is one step in the string of actions developed for that specific scenario. It is up to the retailers to determine what message the customer will receive with each click of the mouse. Relevant, consistent and information-rich interactions with the customer are all parts of what is known as scenarios.

> **Relevant, consistent and information-rich interactions are necessary for a scenario to achieve its desired outcome.**

Automation is a key factor that pushes a scenario to perform its function and desired outcomes. Retailers also use scenarios not only for popular products but also for discontinued and refurbished products. Multiple tasks within a scenario often take place across web sessions and are used to extend the longevity of your customer relationships. Many vendors in the market today offer innovative and technologically advanced scenarios to ensure successful eCommerce fulfillment.

1.3.2 Marketing and Scenarios

The common online shopper probably does not know that they are part of a scenario as they shop or browse your site. Though there is no difference between scenarios applied to retail and non-retail industries, you should be careful to address the goal here—driving sales. Thus, in a retail site, scenarios provide a rich set of interactions to deliver specific messages to customers through marketing, selling and service via web, email, or chat.

eCommerce Best Practices

The goal is to get customers to complete a purchase, whether that process spans days, weeks or even months. Scenarios, to put it simply, are the different messages the retailer decides to give a specific customer, based on his or her shopping click-thrus, in attempt to ultimately guide them to the checkout. This all depends on your absorption of the customers' history and needs. Marketers can use scenario technology to manage a variety of customer experiences through modules like up-sell, cross-sell, email campaigns, product/service promotions, and loyalty programs.

Since "versatile" is the word used to describe the technology, scenarios can be applied to other diverse business needs which may include:

> **Scenarios automate the messages shown to customers during their site visit to maximize conversion rates.**

- Supporting business applications like customer relationship management or supply chain management.
- Building and supporting relationships with business partners, consumers, and distributors.
- Building consistency across all channels.
- Allowing customers to be a participant in more than one scenario.

Scenario technology today allows you to define and evolve the customer experience without extensive IT support. The KPIs are easily captured and measured effectively to determine the ROI. Today retailers do not limit their definitions or applications of scenarios based on what they feel are or are not technically feasible.

1.3.3 Advantages of Scenarios

For retailers, successful scenario technology provides constant demand for rapid returns on investment. They are models which allow retailers to get to the point of competitive advantage. Some of the advantages for retailers to have scenario technology or marketing strategies are:

Increase in Sales

Customer behavior, shopping patterns, history of accounts, demographic data all help to create individualized customer profiles. The more accurate the data,

Operations Cost Reduction

New scenario technology allows full automation on both front-end and back-end processes. This automation helps reduce the amount of labor, IT and customer acquisition costs and enhances operating efficiencies. These technologies are easy to setup and can be absorbed easily to achieve ROI quickly.

Customer Satisfaction and Loyalty

By using the customer history and preferences, it allows you to give them the information, product and services they need without wasting their time. Customers prefer to do business with retailers who give them the most gratifying, relevant and consistent customer experience. Scenarios enable retailers to achieve this loyalty.

1.3.4   What to Look for in Scenario Technology

With the market growth and vendors offering technologies with a scenario module, it is imperative that some sets of criteria be established before you invest in the technology.

Should Be Scalable

The scenario technology should be scalable to adapt accordingly to audiences of any size across different demographics from a single platform. This is useful when your scenario engine has to take care of gathering data from promotions laid out to sales from different markets and age groups or any other classification you want to focus your attention to.

> **Scenario engines should be scalable, and easily integrated with third party applications.**

Ownership/Open Platforms

Linking information from multiple databases should be easy to obtain and maintain. Integration to third party applications should be seamless and customizable with provisions for adding new functionalities.

Proven Technology

Always ensure that the scenario technology is proven and tested across a wide variety of business applications. This makes the technology versatile and compatible for future business strategies.

Scenarios make it easier for retailers to automate messaging and promotions on the website directly to each visitor. Its smart technology that to an extent infers the customer's next step based on their previous steps.

eCommerce Best Practices

1.4 Promotions

- 1.4.1 Introduction
- 1.4.2 Strategies that Make Differences to Promotions
- 1.4.3 Some Quick Ideas to Increase Revenues
- 1.4.4 Some Low Cost Promotional Ideas
- 1.4.5 Promotional Types

1.4.1 Introduction

E-promotions have reached a point where virtually all business categories utilize them in order to drive traffic and sales. Retail giants like Wal-Mart, Kmart and Target are building brands across online promotions using e-coupons or sample sales tactics to increase customer awareness and purchase commitments.

> **Promotions are not new concepts in eCommerce, however, you can still differentiate yourself by using various combinations of promotional types.**

The reality is customers have so many choices when they go shopping and are exercising their options very aggressively. There are numerous types of promotions and combinations of promotions out there, so how do you leverage them in your marketing strategy?

1.4.2 Strategies that Make a Difference in Promotions

Internet Kiosk

"Mini-stores" solely dedicated to promoting your brand and products are great tools to introduce people to what you offer and maintain brand awareness. They generally spread across areas of public interest that allows for a greater reach. Because the offline media still relates with the POP displays, manufacturers are able to have personalized content with a deeper breadth of product offerings. Convenient customer service powered by CRM technologies and readily available product catalogs are more proactive methods of attaining differentiation.

eCommerce Best Practices

Personalized Web Coupons

Coupons are great tactics in promoting products or specific lines of products. Email marketing has come a long way in allowing customers to obtain discounts and coupons, whereas before they would have to wait for snail mail. Coupons generate immediate interest and can yield a greater number of transactions.

Value-added Web Content

With enhanced retailer relationships and educating customers through 'edutainment', retailers are capable of increasing their web presence through various inexpensive or high investment techniques.

Online sampling (emails)

There is no such thing as too much information. Take advantage of the opportunity to sample consumers about their online experience, but be sure to reward them with special promotions or bonus points to loyalty programs just for participating. Consumer research is invaluable to long-term eCommerce growth.

Micro-sites

Micro-sites provide retailers the opportunity to selectively target visitors with specific interests. Generally, micro-sites are specialized "break-away" sites relate to the main site, but is more dedicated to a particular topic. Retailers tend to gain additional revenue premium that does not exist in their main physical environment.

Opt-In Emails

Similar to direct mail, retailers have more to gain by sending opt-in emails to consumers. This can also include co-branded promotions with other organizations related to your campaign. This will give retailers an upper hand in creating new revenue streams from existing customers and can also have targeted marketing segment opportunities.

1.4.3   Some Quick Ideas for Increased Revenues

Usually most retailers are in a bind when deciding what kind of year-end promotions they should run in order to achieve their goals quickly and effectively. Always think about your online business in broader terms. Consider teaming up with other companies to offer a consolidated promotion, for example an airline teaming up with a sun-block company to give away vacation packages to Hawaii.

Some of the other ways to achieve an increase in revenue are:

Include RSS feeds to market promotions
RSS, Really Simple Syndication, feeds are real-time news aggregators that can inform users about specific topics of interest. Make your RSS feeds more interesting with a mix of special offers and product focused information. All of these promotions should be supported throughout your site and in footer links as well.

Send sponsored e-greetings
Go beyond your marketing plan and send out e-greetings for regular holidays like Thanksgiving, Christmas and New Years, but also separate yourself and send e-greetings on more obscure holidays throughout the year. Promotional items that add value to that season could also be added in this message.

Special coupon to use after January 1st
So how do you provide your customers a reason to come back after the hectic Christmas and New Year sales? Give them a coupon for use after January 1st and position the marketing message in a subtle and inviting way, like "We think you deserve a gift for the New Year" or "Why stop now?" This not only makes the customer feel special, but also gives them an excuse to shop.

> **Track your promotions with customer purchase history, product sales, new user visits, and shopping cart abandonment. This data will help you improve the next promotional campaign.**

1.4.4   Some Low Cost Promotional Ideas

Special offer for new users
Any new user who is registering to your site and uses the opt-in email confirmation is entitled to a special offer. This could be contained within a promotion or a coupon. This could also be coupled with a related up-sell offer in the order's confirmation email.

Add promotions to thank you emails

eCommerce Best Practices

Adding special promotions to thank you emails is a novel way to give them another reason to revisit the site.

Employee promotions

Your employees can also send special offers to their friends and family lists. This is useful when you have overstocked items after the sale season. Your employees could be given gifts in the form of coupons or points which is based on the number of family members or friends they can manage to bring onboard and use the purchase process. Your employees are consumers too, so do not exempt them from using any promotional techniques.

1.4.5   Promotional Types

The types of promotions you choose to use will be based on your marketing plan and business strategies. Does one type of promotion work in all retail situations? You will need to test different types and combinations of promotions in order to know what works best for your company.

Special prices type

This is when a specific price or percentage amount is discounted for either a certain product or the whole order. Also, labeling the discount as a "limited time offer" can foster a sense of purchase urgency that would not be there if that phrase was not shown.

Item pricing type

These promotions are fixed on the pricing aspect where there are variations in the percent off or a certain amount off. For example, get X product for 30% off, or get Y product for 30$ off.

Quantity price type

This is the most common promotional type—where you find products based on combinations like buy one and get one free, buy 2 and get one free, or buy 2 of X product and get the third for 30% off. The variations can be made to any combination of numbers depending on your strategies.

Shipping price type

Promotions based on the shipping process of your cart. Free shipping is also a big factor in conversion rates—typically retailers set a minimum pre-tax dollar amount for the customer to meet in order to be eligible for free shipping.

eCommerce Best Practices

Order pricing type
These are based on discounts offered on total price of your order or a free X product if your total order is over Y amount. With the above types you need to know that promotions are first classified based on the target audience and then the appropriate marketing techniques are applied. Marketers often make the mistake of looking at the techniques of what type of promotions to use and then estimate the target audience instead of the other way around.

Free samples type
Various vendors offer 3-15 free sample promotions, where the customer can choose which products they want a trial version of. This is great for customers because they do not have to buy the regular sized product if they just want to try it out and it is great for the retailer because samples drive more future sales on the site. Free samples are good for beauty products because they are generally easy to package in a trial size versus other retail products.

On the shopping cart page, there you should incorporate a promotional code section, in which customers can type in a designated number that associates with a special promotion. This is how promotions can be enforced and regulated. Even though you may have eight valid promotions running simultaneously, typically retailers will only allow the customer to use one promotional code.

eCommerce Best Practices

1.5 Rich Media

1.5.1 Introduction
1.5.2 Does Your Site Need RIA?
1.5.3 Opportunities with RIA
1.5.4 Advantages of RIA
1.5.5 Disadvantages of RIA
1.5.6 Choosing an RIA Solution

1.5.1 Introduction

In order to attain a richer user experience, many retailers have crossed over to Rich Internet Application (RIA) technologies instead of relying on the more conventional HTMLs. RIAs deliver compelling user experiences with interactivity, responsiveness, and richness because they act like desktop applications, which allows for the immediacy of displaying information without an action trigger or link. Put simply, RIAs are tools that are pushing eCommerce into its next generation. RIAs offer different and unique ways at approaching your online store in terms of interface and click-thru processes. They are changing the traditional layout of eCommerce sites and flipping it on its head. Because RIAs add graphical richness, they give your products a highly detailed, interactive, and often eye-popping look and feel.

> **If you decide to use RIAs, try to attain a balance with HTML. Plain text can be displayed by HTML but other products might benefit more with RIAs.**

AJAX, Flash and Flex are three of the most commonly used tools for creating RIAs. Whether it is employing a function that allows you to drag a product image into the checkout box (drag & drop), moving sliders on the navigation bar to select your budget range, or creating a movie clip showcasing the latest fashion line, RIAs are differentiating themselves from conventional websites. RIAs take users beyond the capabilities of familiar page-after-page navigation and present them with an enhanced online shopping experience.

1.5.2 Does Your Site Need RIA?

It is important to separate what you need from RIAs from what you want because even if the technology is the first of its kind, the ROI might not be noticeable. You need to be sure that an investment in RIAs matches your customers shopping behavior. RIAs could very well turn people off when they visit your site because it is too hard to adapt to. In order to consider whether your retail site requires the implementation of RIA technologies, you need to consider whether the deployment of such a technology would lead to greater flexibility for upgrades in the future and lower overall costs of ownership.

Some companies today still fail to meet the basic standard levels of user experience, so when you make a decision to upgrade using RIAs you may want to make sure you have already tried out all the standard methods for your eCommerce system. Keep in mind that RIA technologies should garner greater attention and, thus, thorough testing to ensure that they work properly.

However, the impact RIAs can make on your site can be incredible. They are able to make your website stand out, particularly from your customers' point of view. If developed properly and targeted to the right audience, customers will find your site more engaging and interactive. This will certainly lead to an increase in visits to your site, and ultimately increase your conversion rates.

1.5.3 Opportunities with RIA

Adobe's Flash and Flex, as well as the popular AJAX (*Asynchronous JavaScript and XML*) bring new applications to the web and aim to simplify processes so that users spend more time browsing and, thus, shopping for more products. RIA provides more opportunities to a retailer that is open to change and expanding their customer responsiveness. Some of the challenges and opportunities are:

> AJAX, Flash and Flex are the common technology tools that develop RIAs. They are able to enhance your user experience, making your site more dynamic.

User Experience

Retailers can put a 3D virtual space where the consumer can seamlessly zoom and rotate a product just like seeing it at an offline showroom. Customers get rich brand experiences also from audio and video if, for example, you wanted to show a demonstration of a product/service. You can take advantage of this and ensure maximum exposure of your brand. Sliders, drag and drops, virtual tape measurers, and roll-overs that make product information and selection accessible on one page are other features that can potentially enhance the shopping experience and more importantly increase conversion rates.

eCommerce Best Practices

There are new state of the art Shopping cart technologies that are pushing the shopping experience to the next level. For example, a clothing line can create an interactive mini-movie to be displayed on product pages, in which the customer can simultaneously click on the shirt the model is wearing and drag it to a shopping cart drop box—all while the movie is running.

Responsive Purchase Process

With the advent of RIA, customers have the flexibility of purchasing without limitations. At any time, they have the option to seamlessly go back and forth through their purchase process, remove, update, save and add—all within the same window sometimes. These fluid experiences bring the brand to life.

New Navigation Tools

Customers today are presented with new navigation tools in RIA technology - tools that simplify usage and are highly interactive and visually compelling. Customizable pages, movable interactive icons, dynamic breadcrumbs, and cross-disciplinary searches are just some of the aspects. These tools not only get more people to visit your site, but they end up staying longer, which gives them more time to shop.

Real-time Interaction

With RIA, Web 2.0 and Social Networking, customers can participate in real-time interaction with registered and anonymous shoppers before making their purchase decision. Customers are influenced by other customers, so become more pro-active in your messaging and the ways you can interact with your customers.

Widgets & Gadgets

Creating a widget or gadget GUI tool for users is a great way to engage key target audiences. These tools are designed to make information about your brand or an aspect of your brand readily accessible to your users. Generally speaking, widgets and gadgets show up either on your dashboard or as a desktop feature that can be accessed with just a click of a button. This is a convenient way for customers to readily interact with your brand.

1.5.4 Advantages of RIA

Sorting and Presenting Large Data

Ease of use to sort and select multiple product offerings is more interactive and dynamic with RIA technologies. Traditional methods are time consuming and not as user friendly as RIA technologies can make them.

Reduces Page Reloading Time

Users are able to select and preview their choices on the same screen. One of the biggest advantages of AJAX is that it has the capability of reloading only a designated portion of a page instead of having to reload an entire page. Different images or promotions can be displayed without interrupting the customer's shopping flow.

> RIAs allow for a more streamlined shopping experience because if a customer wanted to make a small change to your shopping cart only a portion of a page needs to be reloaded as opposed to reloading the entire page.

Image Manipulation

Retailers have the advantage of presenting their products dynamically, enabling them to create the same feeling online as if they were physically shopping at their store. Products can be presented, rotated, zoomed in or out, and customized, all in the same page and with only a click of a button.

More Information

Complex products can have prices appearing with user input. Retailers can present products which after user selection, displays the cost, discounts, savings etc.

Simple Controls

All the features that the user gets are made available through simple controls. Thus, simplicity and ease of use with this technology makes the whole experience interactive, which means a greater ROI.

1.5.5 Disadvantages of RIA

RIAs May Be Too Unfamiliar

Depending on how you choose to use RIAs, they may become too distracting or inhibiting during the customer's shopping experience. Because some forms of RIAs are too unfamiliar than what the customer has been used to, they might not notice what you want them to see or do what you want them to do. It is always important to go over your click-thru analysis, monitor conversion rates, and whether the RIA attracted more visitors to the site.

System Requirements

RIAs face an obstacle in whether or not their audience has the system and application requirements they need to support them. If users are not familiar with these features, then they may not be compelled to click or download another application, especially if they are not familiar with it. Flash players are fairly common in mature markets; however, people may not feel secure in downloading a new plug-in. In order to play an RIA made in Flex, customers would need the latest Flash 9.0 plug-in, which even less people currently have. It is also important to make sure the RIAs you use will work across different Web browsers and their various versions.

HTML and RIA Balance

A balance of HTML and RIA should be maintained. One might not have plain text in RIA as it is more logical to have that in HTML, as some find that the editing process is much simpler. It will take some time before RIAs are accepted by the broader market and customers feel completely comfortable with seeing and interacting with RIA technology Retailers should otherwise create a balance where RIA intermingles with plain text.

SEO

Search Engine Optimization is an integral role in online Marketing, so when considering using RIAs be aware that this could affect your SEO presence. For example, although Google has made tremendous strides in making Flash more readable to search spiders, the crippling effect RIA has on SEO is a main deterrent from deciding to use them. AJAX is still ineffective in terms of SEO, which certainly should affect your decision to use this technology.

1.5.6 Choosing an RIA Solution

One factor to consider when deciding which RIA technology to use (AJAX, Flex, Flash, etc.) is the scalability of what infrastructure is required and how much you expect your online usage to grow. Many small vendors are tempted to compete with larger companies, but do not make hasty investments in technologies you do not need. Smaller vendors are quick to use static HTML pages; whereas incorporating RIAs would transform the site into a very dynamic environment. Taking that leap is a big step and one that may not be warranted.

Programming language is also a big consideration. Not every developer is familiar with AJAX or Flex so if you want to create or maintain functions internally it might take extensive searching or training to do so. At least in terms of AJAX, no backend developing is necessarily needed, however for Flex, it is better to have a backend developer that also understands the data and GUI aspects. The learning curve for Flex is significantly steeper than AJAX because of its familiarity in the development community; it basically consists of building upon or enhancing pre-existing HTML based applications. On the other hand, the development time for Flex is shorter than AJAX because it only requires less programming lines and compiles quicker.

Because of the way AJAX is developed, it would benefit you the most if you wanted to change your navigational features or functionalities like drag and drop or sliders; whereas, if you wanted to build a stand-alone, complex application, Flex would offer more rich and flexible capabilities. Flex and Flash tend be geared towards more graphically based applications, strong for creating video clips and animation.

Keep in mind that you do not have to use one solution over the other. In fact, Adobe has a program that focuses on a Flex-AJAX bridge, which entails the interoperability of both into one system. RIAs are flexible, meaning they are not mutually exclusive in nature. Use them to your advantage, but do not overwhelm yourself with the latest and greatest tools. A small change can have an astounding impact on conversion rates, so focus on the big picture instead of getting caught up in the bright lights of RIA technology.

eCommerce Best Practices

1.6 Branding

1.6.1 Introduction
1.6.2 Branding for Internet Retailers
1.6.3 Dimensions of Branding – Offline vs Online
1.6.4 Brand Personality
1.6.5 Online Commercials
1.6.6 Reach Out to Your Customers

1.6.1 Introduction

In a world where companies are trying to enter and, hopefully, play a leading role in the market place, brand equity becomes more important in separating you from the competition. This day and age, a product is only as good as its brand strength and market recognition. Strong brands sell and because companies took the time to develop who they are and who their customers are.

Branding is not just logos, celebrity endorsements, and flashy ads. Although it may end up implementing some or all of those tactics, branding is a representation of your company, the type of products you sell, and the people that buy them. So when you design your logo, hire a celebrity spokesperson, or create an ad, you need to ask yourself whether they are the best representatives of the characteristics you value as a company.

> You brand is all encompassing. It's not just about logos, slogans, or flashy commercials. It's about your company's identity and values.

The power of branding is undeniable. Every image or message your brand puts forth is vulnerable to consumer criticism. If they like what they see, or more importantly, if they connect to your brand image, they will want to buy your product simply because it reminds them of themselves.

Strong brands are not necessarily global or big name brands. Strong brands are those with a focused objective and faithful customers.

Let's take, for example, the iPod™—Apple's holy grail of a portable media player. This product was launched in October, 2001, however, its sales really kicked off three years later in 2004. Brand marketing, in combination with smart product design and the music software iTunes™, made this little media player a cultural phenomenon.

> **The more you understand your core target consumer (values, lifestyle, shopping behavior, etc.) the more you will understand the key characteristics of your brand.**

It is no surprise Apple chose to target a younger demographic for its music/media player—the MTV/millennial generation that seem to have an endless budget, or at least access to their parents' wallets. Apple targeted this group with creative ads showcasing dancing silhouettes, celebrity endorsements like U2 and Eminem, and fresh music by undiscovered artists. The ads set the tone that iPod™ is hip, fresh, and fun—values that young-adults connect with or want to connect with. And what place has the most concentrated population of hip, young-adults? College campuses.

In 2001, you probably would have only seen a couple early adopters with iPods™. Fast-forward to today and it would be hard to find a college student not walking around with the trademark white chord and headphones in their ears. As students saw more and more students carrying iPods™, they realized it was more than just a music player. It represented the "in-crowd", so naturally students were compelled to buy one just to fit in. Eventually, the behavior and attitude within the college microcosm filtered into their parents' and siblings' livelihoods, thus creating an iPod™ phenomenon that spans age, race, and gender demographics. Apple did not just have the answer to a need; they created the need.

One product was able to catapult Apple's brand to a whole new stratosphere. All of a sudden Apple was exposed to a younger demographic, and could build loyalty at an early age.

A lot of Apple's sales growth has to do with the physical store Apple created to purchase Apple products. But what about branding for retailers exclusive to the online world? Do the same branding methods apply there as well, especially if you sell multiple branded products from your site?

1.6.2 Branding for Internet Retailers

The nature of Internet retailing lends itself to rely on rich consumer experiences. Depending on how good or bad the consumers' shopping

experience, the perception of your brand is constantly being formulated. Good consumer experience translates to site loyalty and repeat visits and purchases.

The philosophy behind product branding does not change in an online space, but the way branding is approached is slightly different. To understand this, you must understand why people participate in online shopping instead of going to their nearest store and purchasing the product there. One reason might be that their computer is closer than the store; therefore it is more convenient for them to click a mouse than get in their car. A second reason being they are looking for items that are not typically found in most stores and online vendors more readily cater to niche markets. Another reason, possibly the most popular reason, could be the fact that they are just searching for the best deal.

> An eCommerce site's branding success depends on the quality of the user experience. Make it multi-sensory and provide extensive product details and images.

There are different ways a retailer can brand their company. User interface, navigation, and site design are some of the most important factors that form your brand. It instantly builds credibility and reassures the consumer that you are legit. Unlike shopping at a retail store, customers cannot touch or inspect the products they are considering buying, which to the customer is a potential risk when online shopping. You want to reduce this risk by creating a sound and reliable online shopping experience that reflects your brand and customers.

First Impressions are Lasting Impressions

The worst case scenario would be to have a customer stumble upon your site, look around for a couple of seconds and revert back to the search page for another option. First impressions of a site should be advantageous to you because you have the power to design a professional looking site. You can control what the customer sees or does not see. Simply having an online presence is not enough to have any long-term stability.

Great Expectations

Another scenario that is just as bad is if the customer directly goes to your site, looks around, only to be let down. This customer expected to have a better experience on your site than you actually delivered. You do not want to disappoint potential customers, so it is best you start trying to understand your customers. Why do they visit your site and what for? Of the people that visit your site, how many actually complete the checkout process? By getting to know them, you can create an experience that best fits their needs and start building your brand's identity.

Best Case Scenario

The ideal situation involves customers either stumbling upon or directly going to your site with a specific product already in mind. Your site is built to perfection and they immediately find what they were looking for, and because your site was personalized, they find additional items that they did not even know they wanted or needed. They not only purchased the item they intended to buy, but they added on products during their shopping experience. This is the sign of a quality site—one that gives customers a great, worry-free user experience.

The quality of user experience essentially makes or breaks your brand. Because so many customers are online shopping, it is important to capture the essence of your brand onto your computer screen.

1.6.3   Dimensions of Branding—Offline vs. Online

Market penetration is difficult for up and coming companies looking to find their portion of the market share. What essentially separates the boys from the men is that X-factor; the branding gold that is able to hit people's sweet spot. The problem with branding today is that the everyday consumer is exposed to so many brands and has so many avenues to reach those brands, online and offline. It is up to you to actively differentiate yourselves from the competition, and take a clear position as to what your brand represents.

> **Keep the retail store and online store consistent. You want the customer to recognize your brand when entering the store or site. Harmonize logos, copy, images, models, music, and colors.**

For those brands that have physical retail stores as well as an online store, branding has to be consistent on both playing fields. The jump from the retail store to the online store should be seamless in design and feel.

Environment Consistency

When the customer goes to your site, he or she should know immediately that they are experiencing the same brand as in your retail store. This can be done by not only putting your logo and the

same color schemes on your site, but taking it a step further by mimicking the images in your retail store—photographs or artwork, using the same music in your stores to create the same atmosphere, or choosing the same models to showcase your clothes.

Just like how the retail store is organized, the online store should be organized in the same manner. For a clothing store, the online store should offer the same separated departments, like menswear, women's clothing, or kids, and from there separate it by the new arrivals section, T-shirt section, pants section, sale section, etc. The entrance of the offline stores should be a reference to how the homepage of your site is designed. If there is a big sale sign in front of the retail store you might consider that your homepage needs to promote that same message.

1.6.4 Brand Personality

As it has been stated earlier, much of your success has to do with the intangibles as opposed to the actual products you sell. So how do you create your brand's personality? First you must look at whom you sell it to.

Who Are You?

The best way to discover the essence of your brand is to discover the characteristics, lifestyle, and values of your target consumers. Find out what is important to them, not in a product but in life, and you will soon discover what is important to your brand's identity. The more specific you get, the better your understanding you will have of your brand.

Think outside the box and humanize your brand by creating a story or biography about it (not your company history) including key turning points in your brand's "life" and how it became the "person" it is today. This might sound like a juvenile exercise, but it will give you a well rounded idea of who you are as a brand.

Company culture should also be put on high importance because it allows you to evaluate the things that identifies and maybe separates you as a business. If you are focusing your efforts and spending a large percentage of your budget on providing the best customer service out there, then it is highly likely that the people you hire on staff will reflect this attitude, which then of course affects how they relate with customers. The standards you hold up as a company will make its way to your customers, and based on their perception they will ultimately make a choice in whether to continue to do business with you or not.

Looks Matter

Once you have developed an understanding of your online brand value and consumer, you then can create the physical manifestation of it—logo, colors, font, symbols, slogan, commercials, print ads, celebrity endorsements, etc. Figuring out your brand's personality was fundamentally the hardest, most crucial part and everything else that follows is a manifestation of those qualities.

Price and Promotions

Promotions play a significant role in online marketing, as they are easy to send out to registered customers and are effective in creating brand loyalty, or at least repeat visits. What can your brand say or offer to your consumers to achieve maximum sales potential? Regular promotions or an annual event sale are some things customers look forward to, sometimes even wait for.

Be a Social Butterfly

Because of the social networks, blogs, and viral media available to consumers today, your brand, and subsequently you brand personality, gets exposed to more people than ever before. This can play to your advantage in many ways, as it lends itself as a forum for praise in a way that reaches millions of potential customers. However, it can play against you as well because it is at risk for criticism that could be harmful to your reputation.

1.6.5 Online Commercials

The presentation capabilities of the Internet can only enhance your brand, so why not use it to your advantage? Not only can you use RIAs, Rich Internet Applications, to make your eCommerce site more interactive, but you also have greater creative avenues of marketing your brand. Television advertisements are being threatened by technologies like TiVo and DVR that allow viewers to fast forward through commercials. Companies are spending large amounts of advertising dollars with television programs, but what is the point if nobody sees them? Creating online commercials or putting commercials on either your eCommerce site or other sites is a great way to get exposure, especially because of the relatively easy accessibility of the Internet.

eCommerce Best Practices

Some companies like American Express and BMW have invested in creating commercials specifically for online purposes. There are a few motives for creating an online commercial versus television commercials. One is that it will increase traffic to your site, thereby exposing more people to your brand. Another is that some companies request that viewers register with their site, thus obtaining more user profiles and information. Also, if viewers of the commercial find it entertaining or relevant, they are more likely to peruse your site after they have finished watching the ad, and at the very least walk away with a favorable perception of your brand.

Everything from watching the commercial to making a purchase can be done in one sitting with online ads, whereas there is a greater disconnect and time-lapse between watching television ads and actually buying the product.

In addition to professionally made commercials, avenues like YouTube are giving the public a platform to express themselves through video. Some consumers that are loyal and proud of certain products and brands have even gone so far to create their own commercial or mVid, music video, about them.

1.6.6 Reach Out to Your Customers

Take advantage of the web and e-mail because they offer new avenues to reach your consumer in more ways than sending out messages. Here are some ways to leverage the medium in your favor:

- As mentioned above, social networks, blogs, and viral video have an increased presence online and can be used to enhance your brand. You would want to visit sites that serve your brand's interest, for example if you sold snowboarding products you might want to take a look at certain ski resort forums or join a University's snowboarding club social network. They offer ways to extract information from consumers about their likes, dislikes, concerns, wishes or they can serve as a way send out promotions to targeted groups. As for viral videos, the snowboarding brand can post monthly videos of tricks or wipe outs online.

- E-mail is a way to keep your consumers connected to your brand. Send them updates, special offers, or messages to make sure you are top-of-mind.

- The more interaction the better. Plan events that gather consumers together and promote your product in a creative way. Use your online capabilities like your website or blogs to get the word out and use your retail stores as a primary location of the event. If it is a success, you will hear about it online the next day. The web can give you immediate feedback and consumer honesty—sometimes brutal honesty.

The formation of your brand can make or break you. It is not solely about the products you sell, but the values you stand for as a brand. Ultimately, consumers see through the smoke and mirrors, which has lead to the demise of many companies.

Branding in the online world should only work to your advantage because you can control what the consumer sees and experiences. There should be no excuse for customers to abandon your site without even taking a look at the products you sell, but it happens because of weak user experiences. In the online world, poor user experience equals poor branding.

Success starts with discovering your brand's online personality and bringing it to life.

eCommerce Best Practices

1.7 Social Networking

1.7.1 Introduction
1.7.2 Popular Mediums
1.7.3 What to Consider When Interacting with a Community
1.7.4 How to Convert Browser to Buyer in Social Networks
1.7.5 Analysis

1.7.1 Introduction

The concept of social networking has been here for a very long time albeit not necessarily in the context of the online world. Online social networks are the links to retailers and their customers—they are ways retailers can get closer to their consumers.

We live in a society where we need contact with people, as such to create and maintain a balance within a community known as a social network. Communities within the social network assume many shapes in terms of communication—these can be blogs, wikis, user generated content like YouTube, product review sites like download.com or even eCommerce sites like Amazon or eBay. All have a common purpose—to facilitate and encourage social interaction between users with the same interests.

How can you tap into these social networks and reach out to those communities? Or can you create your own community? MySpace.com, Facebook, and YouTube are some of the most popular headliners in social networking. Facebook and MySpace allow people to create their own "homepage", link to other registered members, and create or join groups with like-minded interests. Groups or communities can range from the more straight-forward, mass appeal type to the more obscure, niche type. Depending on your brand or your business goals, there might be a group your company can be a part of.

> **Know who you want to target before entering a particular social network. Present your brand in an interactive way so that you get positive feedback, but be prepared for negative feedback—it is part of the risk you take in social network marketing.**

Segmentation is typically not an issue in social networks because members tend to group themselves into interest-based segments as they become part of a network. Shared interest is vital and retailers should not target these users for segmentation but rather their interests. This is known as behavioral targeting. Audience reach and user engagement are some important factors in social networking.

Since social networks are places where people share opinions, insight, experiences and perspective with each other, there are different forms it may take like text, images, video and audio. Most popular mediums include Blogs, Message Boards, Podcasts, Wikis and Vlogs.

1.7.2   Popular Mediums

Blogs: This is a user-generated website where entries made in journal style are displayed in a reverse chronological order. They often also provide commentary, news, or opinions on a particular subject which could range from any number of things like food, politics, fashion, celebrities, and of course, marketing.

Message Board: Similar to Internet forums, the functionality is similar to blogs and is much older than the former. People generally showcase their opinions on certain topics/threads that are of particular interest to them and other users will then comment or ask questions related to the topic.

Podcast: This is a media file that is mostly distributed by a paid or unpaid subscription over the Internet using syndication feeds like RSS. The video/audio files are typically played over PDAs or personal computers. Podcast is derived from Apple's iPod, where the files were intended to be played.

> **An honest description of your brand has to be communicated in order to experience any long term loyalty. Communities are sensitive to certain messages and the only way to protect yourself is being upfront with members.**

Wikis: These are websites where any user, registered or not, can add, edit, or remove content to the site. Ease of authoring and operations for such a mass scale method make it popular.

Vlogs: Also called a Videoblog, this combines video and blogs. Authored by individuals, Vlogs take advantage of RSS or Atom for distribution of video over the Internet.

1.7.3 What to Consider When Interacting with a Community

Brand Representation

Make sure you are staying true to your brand's identity whenever you interact with your consumers or audience. The language and graphics you put forth should be consistent with your brand because that is what your consumers identify with right off the bat. How you engage with community members needs to be in keeping with who you are as a company.

Don't Be too Aggressive

Consumers within social networks have their own independence and have joined at their own will. You should also embrace that—do not barge in and make your presence overbearing in any way. Instead you should ease your way into their community and observe for a while before you start to interact. Once you become a part of the social network, your marketing strategies should not be pushy. You have to respect and understand why the community was formed in the first places and then target your efforts to run parallel with the likes and dislikes of the consumers.

> **Always engage in a dialogue when you are new to a social network. Patience is a virtue—don't just advertise your brand because it might not sit well with community members. Spark interest by engaging with them.**

Have Valued Input

It is not enough to just be part of a community to get the desired results. After claiming your spot within the network, your next step would be to create a valued role where consumers look to you as an expert. It will be to your own discretion whether you want to divulge in the beginning that you are linked to your company or not—some members may appreciate your candor. The facts and viewpoints you give them, though they might not believe everything, will be used as a communication point for information about your company.

Listen Carefully

If you are able to get a collective group of interested customers within the network, have a plan in place that can collect their feedback about their experience with your brand. It is not only a way to interact with them, but it also solicits their feedback which should be an integral part of your marketing strategy.

Be Selective and Prepared

Social networks are largely uncensored communities. When you allow your brand to enter that world, be prepared to expose your company to questionable content in which there is limited monitoring or control. However, a social network is a great tool for damage control because it generally is an isolated community in which its content can be relatively contained.

Never Advertise; Engage

Instead of explicitly advertising on the networks, make the other members feel special for being part of the community. Try to offer something special that the consumers cannot find elsewhere, something unique, for example a special deal on a car, a different school loan offer, tickets to a consumer's favorite band or show. People are more likely to appreciate your brand when they can associate it with a pleasant personal experience.

> **Reward community members with exclusive promotions. This might incite a viral effect, which ultimately leads to greater brand recognition.**

1.7.4 How to Convert Browsers to Buyers in Social Network

It is a difficult task to find out whether 'x' number of accounts in a social network space are all individual ones—some could be while others could be duplicates of the same individual. You need to adopt a sharper profiling system, one that changes with what people feel like at that moment. Users generally have a visual tag to identify them and describe what their mood is at that time, for that day and during that week. Accordingly, people can decide whether it is a suitable time to contact you or not. Looking at current interest indicators, savvy retailers can target consumers based on their mood and availability at that time. It is a difficult task to sift and sort through specific moments, but the advantage is that you know the behavior and feelings of individuals. From this you can infer which individuals would be more likely to convert on different offers or promotions.

Social networks provide a growing opportunity for retailers and there are no laid out rules as to how you become successful. It is one place where you can push and/or pull your brand to individuals whose interests are more relatable to your company.

1.7.5 Analysis

Once your online community is up and running, you should have metrics that measure the impact of the community to your brand and products. Some of the standard metrics which you should consider are:

Signups: Look for the number of signups you have received for the community that can be tracked through newsletter opt-ins.

Refer or Forward messages: Another option to look at is the number of 'send a friend' messages. This shows the number of users who actually are spreading brand awareness through the social networking portal.

Time spent: Most of the consumers spend more time with community content like films and music. Keep a track of the average time spent on these because the longer they spend on a site, the more chances you will have to gauge the consumers' interests.

Conversion rates: Based on the information given to them, consumers might end up in your shopping space and, in some cases, will make a purchase. You should estimate the percentage of community members where conversion rates occur.

Revenue influence: Analyze and track the amount of revenue contributed by community members with your marketing strategies. This can be tracked if a specific promotional code is given through the social network channel alone.

User's references: Track the number of friends added to the social network circle with reference to any new marketing initiative.

Social networks are shortening the gap between marketing messages and consumers. Instead of letting their brand slogans do the talking, social networks are now letting retailers provide brand awareness directly to their consumers. The communication dynamic is shifting in what retailers can tell their consumers and how they can say it. Engaging in an open dialogue with your consumers can be considered a risky move, but many will find that it has its benefits because you actually stand to learn something about your consumer and not just the other way around.

eCommerce Best Practices

1.8 Affiliates

1.8.1 Introduction
1.8.2 Benefits to a Retailer
1.8.3 Content is Key to Affiliate Marketing
1.8.4 What to Consider
1.8.5 How to Make Affiliate Marketing Work
1.8.6 Affiliate Tools—Blogs, RSS, Podcasts

1.8.1 Introduction

Today, affiliate marketing is one of the fastest-growing and most important ways to build brand awareness and increase sales. So how does affiliate marketing work? The basic idea behind it is acquiring strength in numbers. You select and make arrangements with other web site owners for becoming your "affiliates." Each of your affiliates put a pre-approved link on their site with a click-thru going to your website. You pay the affiliates on the basis of click-thru, inquiries to your products, etc. Many presume that the more links you have the better off you will be; however, there is also a risk to having affiliates. They can dilute your brand, so carefully decide whether you need to invest in them or not, or regulate the number of affiliates you have.

An affiliate site usually promotes many other brands and products in addition to yours. With that, be privy to the fact that the competing brands could be much greater in awareness than yours. On the other hand, you do get greater exposure and customers will find your brand.

Apart from delivering relevant and organized content to consumers, affiliates can play a vital role in building customer loyalty. The products they promote are not chosen by the affiliates, so they do not have control over the type of products they carry. What they can control is whether or not the products they promote are worthy of being on their site.

> **Affiliate marketing helps increase brand and product exposure because they are shown on multiple sites as opposed to just your eCommerce site.**

The choices of products offered should be relevant and displayed in a way that makes the customers come back. Most of the time financial interests govern the products offered, but if affiliates are able to find the right balance with the

products that are most likely to benefit the customer, they go a long way to ensure customer loyalty with their site and your brand.

1.8.2 Benefits to a Retailer

Affiliate programs can be set to display an advertiser's message to a very select audience. The affiliate program is more like an agreement that you set with a merchant to acquire targeted traffic, generate sales leads and ultimately increasing conversion rates. This is a powerful and cost-effective way to find new prospects and turn them into regular customers.

Individual affiliates also rely on marketing and advertising on their site. Most have an incentive to join up with companies that suit their advertising strategies and can translate their existing audience database in conversion rates. Some affiliates can also be managed in-house as this approach allows marketers the freedom to write customized contracts and create their own rules on how the affiliate marketing program would operate.

You do not only have to pay for exposing your message to prospects; affiliate marketing allows you to set the pricing based on cost-per-click, cost-per-action, cost-per-lead or even cost-per-sale. A number of infrastructure companies also serve the marketers and their affiliates.

> **Prices for affiliate marketing is usually based on the cost per click, cost per sale of the items.**

The bottom line is that affiliates are here to help your brand attain a larger reach than what was established before.

1.8.3 Content is Key to Affiliate Marketing

Though affiliate marketers have multiple techniques to make money, the one thing that would help them along is making sure they publish quality content. So how do you write quality content? Some of the characteristics are:

- <u>Create unique content:</u> Ensure that the content you write is different in presentation and is not more than just re-hashed content that appears on every other site. If the style of writing is sharp, the word choice is relevant, and the presentation is eye-catching, the user will make a point to come back to your site.

- Content has to be user-focused: Users search for information that can help them, especially for information related to tips or how-to articles. As a retailer you need to target them with user-focused web content to which they might keep coming back for more.

- Grammatically correct: Spelling, punctuations, passive voice all add up to well written content. Always omit needless words because being succinct fairs better than long winded explanations.

- Easy to read content: Bulleted lists, short sentences, short paragraphs are some user friendly way to make your content more accessible to online readers.

- Honest content: Ensure that the products carry an honest description and not something that attracts users under a false pretext. It is not a wise idea to promote it as something that does not hold true to its description.

1.8.4 What to Consider

There are a lot of things to pay attention to, and also avoid, when developing an affiliate marketing plan. Affiliate marketing cannot create miracles. It needs support to be successful. Some of the steps to consider are:

Using Viral Marketing Elements

Since affiliate marketing also spreads through word of mouth, viral marketing is equally important as it is an environment where emotional and psychological cues encourage people to drive others to your site.

Affiliates Support

Response has to be quick so that your affiliates feel that their problems are being solved or questions answered. You should ensure that your affiliate programs utilize newsletters, community-building sites, and sales building communications.

Track Performance

All your affiliate programs should be tracked on a frequent basis to receive better results.

Filtering

Since the success of your affiliate program depends on your affiliates, you should ensure that the best fitting, most popular affiliates are held in your program. Filtering out the unproductive affiliates will help you manage what worked and what was unsuccessful.

Tweak and Monitoring

Affiliate marketing often works best when prospects can choose from a variety of products and services that consistently change in appeal and presentation. Each time you make an adjustment, track the results to see who is producing, who is not, and where you can make a change to generate more results. History has a way of repeating itself, so analysis can prevent you from making the same mistake twice.

1.8.5 How to Make Affiliate Marketing Work

Affiliate marketing increases sales when you enlist multiple distribution outlets that have access to your prospects. Preparation is important for the decisions you make and your success in the future. You might be faced with challenges along the way, but take a moment to evaluate the issues at hand. Below are some general points to consider:

Identify and Restrict

You can find a great number of people who sign up for every affiliate program available but do they know exactly who they are? Are they the best choice as an affiliate? What are their characteristics and where can you find them? You need to know that every affiliate you signup for is representing your brand, product and service. It is important that at an early stage you identify your affiliates, restrict your marketing efforts and target them to people who are likely to increase your conversion rates.

Plan to Convince

Affiliates reserve the right to deny your business. Develop a plan to convince top-quality affiliates to sign-up and stay with you. They may be reluctant to switch or take more responsibilities as they might be working with other retailers or even with direct competitors.

eCommerce Best Practices

Structure the Rewards

When signing with successful affiliates consider having a sound rewards structure in place in order to maintain that partnership. This way your affiliates will maximize their efforts.

What to Offer

Web banners, buttons, links, advertisements, landing pages, and shopping carts are some features to offer to your affiliates, but you need to provide both proactive information and quick responses to their inquiries.

Making it Known

Once your affiliate program is tested and implemented, your next effort should be to submit the same to affiliate directories to develop and ensure presence on discussion groups or affiliate marketing newsletters. Also supporting and establishing a community of your affiliates has to be done along with networking with other affiliate marketers. Though the efforts are extreme it is a necessity to be proactive with your affiliates.

1.8.6 Affiliate Tools—Reviews, RSS, Podcasts

Affiliates should also be proactive in driving people not only to their site, but to your products. They, too, can take part in the social media in order to increase traffic to their sites. Reviews, RSS, Podcasts are some tools available to eCommerce sites, as well as their affiliates.

There is no guarantee that they will work for your retail products, but be aware that they are becoming increasingly popular in key demographics. Your marketing strategy, especially when related to an eCommerce initiative, should be open to new ideas and new media. There are many tools out there that can increase awareness to a targeted demographic.
These include:

> **Find out how the affiliates are driving people to their site. If they are not proactively using marketing techniques, it may not be useful to you.**

Reviews

Reviews are full of strong statements and strong opinions. Because of their nature they promote

eCommerce Best Practices

discussions and ideas at a rapid pace that can either work to your advantage or work against you. Unfortunately, you cannot control what is said, but it adds credibility to your company. The worst marketing strategies are the ones that are stagnant. Do not always play a safe game—be open and creative, but also be prepared for criticism.

RSS

RSS on the other hand can engage customers, build brand image, and extend relationships. RSS offers a set of web feed formats that allow you to update web pages frequently. As a retailer you can take advantage of blogs or RSS by offering one new product daily. Being an alternative to email, you can also use RSS to drive time-sensitive information about new sales options to targeted customers.

From a marketing perspective you should always strive to be different. The content you send or display to targeted customers should offer compelling information.

Some eCommerce tactics which could extend your marketing success with the use of RSS as a primary tool are:

- Drive time-sensitive information through RSS as some implementations can direct consumers to the shopping process.
- Most content providers' sites have a provision for advertisements—use them thoughtfully so that it engages users to sign up for RSS alerts about special offers or sales promotions from your site. Updates can also be automatically routed to send messages or news about the brand or product. Also, extend site content with feeds from other sites and blogs.
- Develop cross partnerships with relevant content sites to build traffic. Use the RSS medium to attract a content site's readers to your product due to its timeliness.
- Include widgets that are fun and creative in your RSS feed. Certain products may ask consumers for tips and tricks—your widget can accomplish this.

Podcasts

Podcasts are similar to RSS but the content is in audio form, which the customers can transfer to a compatible device and listen location independent. Though a great marketing tool, it is not guaranteed to fit into every marketing plan. One can extend the brand emotion as a connection to your audience or even facilitate internal marketing communications with your sales, distributors, business partners and employees. If implemented, you should also promote podcasts throughout your retail site by navigation, footer links, email marketing, links to relevant content, etc.

Affiliates exist to support you and your business. The brand exposure you get, or exposure the affiliate gets for that matter, is essential for your brand's growth.

eCommerce Best Practices

1.9 Sweepstakes

1.9.1 Introduction
1.9.2 Sweepstakes Rules
1.9.3 Post Sweepstakes Analysis

1.9.1 Introduction

Sweepstakes are effective marketing techniques because they drive brand recognition and awareness, as well as generate more customer adoptions because everybody wants a chance to win fabulous prizes. A staggering number of unregistered visitors to your site will register once they have an incentive to do so. Sweepstakes have become a part of the marketing strategy because of two reasons—one is that they generate a concentrated list of people interested in something about your brand, and second, they give you an opportunity to gather and analyze data about those people. Not to mention, the winners of the prize will associate your brand with all the good things garnered from their sweepstakes experience.

> **Sweepstakes are still effective in getting people to register to your site and attaining useful customer profiles.**

In retail sweepstakes registrants do not have to give up any major personal information or disclose their credit card details in order to enter. So where do you start? Gathering names and offering a prize does not necessarily get you potential customers. You need to have a planned strategy, know what type of sweepstakes to integrate, the legal issues involved and the opt-in method. All of these, when done in the right way and at the right moment, can provide you consumers who not only participate in your sweepstake but also become a regular customer.

> **If the prizes are related to your campaign / sweepstake's message, it increases opt-in rates. For example, winning a trip to Hawaii for a sun block company.**

There are numerous types of sweepstakes in the market today and they are difficult to proposition to potential customers as the number and types of sweepstakes change with the season and market. However on a general note, there are sweepstakes that are single-entry, multi-entry, instant winning, match and win, collect and win, etc. The type of sweepstakes you choose is at your discretion.

eCommerce Best Practices

1.9.2 Sweepstakes Rules

Retailers should approach each sweepstake as a contract with each and every participant. You should draft a complete set of rules tailored for each sweepstake promotion which should be made available to every participant. There is no standard set of rules that govern each type of sweepstake because each promotion is unique in its presentation and rules. Some of the minimum points which you should address to the participants are:

Purchase or No Purchase Necessary

The participant should know whether he or she needs to purchase a product or not before entering the sweepstake. Clear information on the banner or promotion slot needs to be communicated explicitly.

Sponsor Identification

Sponsor identity, names, and brands should be visible on the sweepstake promotional slot. Sometimes participants get attracted to your sweepstake promotion based on the sponsorship you carry. A smaller retail site can add credibility to their sweepstakes by partnering with known brand names as co-sponsors, and having them put a link to your sweepstakes registration page on their site as well. This will draw more visitors to the site.

Eligibility Requirements

Sweepstake participation can be restricted by age, occupation, residence, or gender. Sponsor employees, agents, members of sponsors should be excluded as their participation may be deemed a conflict of interest. Sweepstakes designed for specific age groups require that information to be visible and vital.

How to Play

Instructions on how to play the sweepstakes should be laid out in a clear and concise manner. Often the language used to describe the rules of the game is more confusing than need be. It is in best practice to be as brief and concise as possible.

 eCommerce Best Practices

Time Span

Every sweepstake should have a starting and ending point, and it is important for every participant to know these dates before entering. Start and end dates should be applicable to online and offline participants to enroll. If the sweepstakes is open to mail service registration, the post mark date is typically used as the marker for accepting further participants.

Prizes

Prizes are also an important factor to get users to enter in your sweepstakes. Prizes can be tied in to your existing product line or possibly a new product that is being launched. The more the types of prizes relate to your messaging the greater the likelihood that the participant will have brand retention. Generally, the bigger the prize the more impact your brand has on participants.

Winning Odds

Certain sweepstakes have a limited number of entrants, and in this case, the rules should state the exact odds of winning. If there is no limit to the number of entrants then the rules should state that the odds of winning depend on number of eligible entries.

Notification and Selection of Winner

Participants are required to give out certain contact information when the sweepstake promotion selects and announces the winner. Registration should clearly state that certain information about the participant is required for eligibility. Also, if you know that the media and/or press will be involved post sweepstakes, then the participants should be privy to the fact that such services would be necessary upon winning.

> **Always capture participant email addresses for future marketing purposes and winner notifications.**

Liability Limitations

Retailers must ensure that all sweepstake rules have been met. The prize winner(s) should sign a document stating that they met all the requirements laid in the rules and regulations. Certain limitation of liabilities could also be included where misinterpreted or incomplete entries are not eligible.

Void Where Permitted

In many countries the sweepstake is strictly prohibited or may involve complex legal issues or even approval from the governments. You should have a clause within the sweepstake rules that allows a degree of protection from these issues. Since online sweepstakes are available to each and every user, demographic legalities should be taken with care and be clearly specified.

1.9.3 Post-Sweepstakes Analysis

Survey

After your sweepstakes are over you should receive feedback from your opt-in customers by having them fill out a short survey. Create an incentive for filling out the surveys, this way participants are still rewarded for participating in the sweepstakes even though they did not win. Gift cards, coupons, or bonus points are great incentive tactics for getting people to voice their opinion about the sweepstakes.

> **Post sweepstake analysis will give you a better idea of how to improve your processes for future sweepstakes.**

Online Analytics

Make sure that a reliable tracking system is implemented prior to the start of your sweepstakes. This makes collection and analysis of your customer data more transparent for future sweepstakes or campaigns. The registration information alone will garner enough data to track whether or not they have returned to your site or not.

Sweepstakes are still highly effective tools in brand promotion and customer profiling. The amount of visitors willing to register for a sweepstakes offering a relatively small prize is still significant enough for data collection. Everyone wants a chance to win something, no matter how big or small.

eCommerce Best Practices

1.10 Newsletter

1.10.1 Introduction
1.10.2 Does Your Company Need Newsletters?
1.10.3 What to Integrate as a Marketer
1.10.4 What to Measure from Newsletter

1.10.1 Introduction

Similar to emails but not to be confused in their delivery and strategy, newsletters are still powerful and effective tools in retail marketing. Compared to emails, newsletters are more time consuming to produce as the content is greater. However, on the recipient side, the medium is sometimes preferred due to its substance and breadth of information. As a medium they can create a more effective bond between the user and your product than with your website. Usually everyone looks at an email as a low cost method to maintain customer contact over time. But newsletters play a larger role than that—it supports additional areas and information that can generate new sales opportunities.

1.10.2 Does Your Company Need Newsletters?

The purpose of the newsletter is to inform people of interest about current or future activities within the company. Whereas email marketing serves to drive quick sales, newsletters serve to maintain customer retention. It is in best practice to send newsletters over the course of the year, albeit less frequently than email marketing messages. Newsletters require more time and effort into producing because it takes as much amount of time to generate significant news about your company. But do not assume the newsletter has to be heavy on the text. Be creative in how you structure and layout the letter because you will surely lose some viewers if it is strictly text. Some general points to consider are:

> **Newsletters are more content heavy than most email marketing techniques. It is important to know who you are writing to in order to create valuable content for the mailing.**

Up-Sell / Cross-Sell

With its larger visual space for content and images, you can cross-sell and up-sell all products within the context of the article. Layout and design are highly important for keeping readers interested.

Less Frequency and More Valuable Content

Every department, of course, cannot send separate emails to all your customers because eventually they would be flooded with numerous emails from just one brand. Use the flexibility of newsletters to compose and integrate different departments' news into one consolidated document. Typically retail newsletters are sent every quarter or twice a year depending on the news and importance of the messaging.

Provide Value

Since a newsletter's primary goal is to build a positive relationship with your customers, make sure the information included in the letter is of value to the people reading it. Understand their interests and structure the news to fit their tastes.

Cross Promote

Newsletters, when designed properly, provide a powerful venue for cross promoting and marketing your additional capabilities or products, as well as sister brands or partners. Whether with text or images, there are many ways to display a newsletter. Use the right layout and messaging and you can achieve targeted cross promotional goals.

Since your email newsletters can lead to new sales opportunities, assume that your customer has a requirement void you can fulfill. Present your services or products in a way that lets them know you have something they want or will want in the future.

1.10.3 What to Include in the Newsletter

Newsletters are more complicated to structure then emails because they can be extremely generalized in terms of topics of discussion. You will want to limit the number of stories included in the mailing, and filter the stories by importance or topics of interest. Another option would be to create a theme for the newsletter and only include stories that relate to that theme.

 eCommerce Best Practices

As important to the general structure of the newsletter are the content, design, and presentation. These three factors play a pivotal role in the initial decision to even read the newsletter. Nobody wants to read a document that looks complicated and text heavy within the first few seconds of opening the mailing. There are some helpful tips and guidelines to writing newsletters, some of which are:

Create a Theme

Giving each newsletter a particular theme will frame the context of the stories. This not only helps you determine what stories to include in the letter, but also lets the readers know the intention of the mailing.

> Create an overall theme for the newsletter and incorporate stories that support it. This gives focus to the mailing.

The Importance of Headlines

Only deliver what the reader would be interested in. Since newsletters offer option of space and less frequency, you should know that content is paramount to delivering brand awareness. Headlines for different stories help the reader decide which article they want to read. Be as clear as possible and only include information that would be valuable to you and the reader.

Consistent Brand Experience

Due to less frequency, brand experience has to have more consistent impact as the previous newsletter. Create a strong newsletter template, both is look and feel, which can be used on a consistent basis to structure your stories around. Some amount of design flexibility should be given in case a particular story needs a change in structure.

> Brand experience and design should be consistent from newsletter to newsletter.

Opt-Out Option

Always give the subscribers an option to opt-out of the mailing list. You do not want to upset a customer by repeatedly sending them unwanted newsletters. Generally the opt-out disclaimer is located outside of the newsletter fields so it is not mistakenly selected.

Always Test with a User Group

Since newsletters incorporate a fair amount of content and graphics, you need to test them before sending them to your customers. Test the copy or headlines in Google Ad Words to see the response and apply what worked in your next newsletter.

Tracking System

With your newsletter having layers of information, you need effective tracking metrics that can record what exactly your customers are clicking. Monitor their behavior to see which sections are most clicked and whether users prefer visuals or headlines to navigate from your newsletter to your site.

1.10.4 What to Measure from Newsletters

Usually the accuracy of tracking metrics is limited by the analytical technology employed. Analysis is the only way to improve upon what has been done before. Below are some characteristics of newsletter distribution that you need to consider and measure:

> **Tracking the opened vs. non-opened newsletters and the click thrus from that newsletter will give you a better idea of its effectiveness.**

Determine the Number of Opened Newsletters

Open rates for your newsletters show how the users responded. You can capture metrics so that you know whether the user has previewed the newsletter on their preview plane or actually opened them up. This lets you know if newsletters are beneficial to your marketing strategy or not.

Click-thru

A click-thru is a sequence of engagements or "clicks" that the user goes through from the links within the newsletter. Your newsletter can be designed in such a way that the user needs to click on a link in order to go to a particular page in your site.

Funnel Navigation

This defines conversions that allow you to watch where your email recipients are going after they received and clicked your newsletter. The

funnel navigation helps you decide whether the promotion strategies were a success or failure due to poor design or unmet expectations.

Conversions

The number of products purchased as a result of your newsletter does not guarantee that the marketing strategy was a success. Registrations, whitepaper downloads which are also linked to your newsletter can be measured. One way is using promotional codes or source codes wherever possible to track these aspects.

Depending on what your ultimate goal is for sending newsletters—brand awareness, conversion rates, etc.—the effectiveness of the newsletters can be measured against it. If you find that they are not worth the time it takes to produce then newsletters might not be one of the Marketing strategies you adopt.

Newsletters can be excellent communication devices between retailer and customer when used appropriately. They offer more prevalent information than an email, which some customer deem important for maintaining loyalty to your brand.

eCommerce Best Practices

selling

ecommerce
bestpractices

eCommerce Best Practices

SELLING

Once traffic to an eCommerce website increases, it is important to translate that into sales growth.

In order to sustain that level of growth, a company has to build an online environment that appeals to their consumers. This does not only mean offering consumers the best products or prices, this means offering them a shopping experience that leads to purchases.

Everything from fraud prevention and wish lists to gift cards and inventory management relates to whether or not a consumer makes a purchase or the extent of his or her purchase. Maximizing sales potential is key to growing business because there are always undiscovered ways of creating loyal customers and repeat buyers.

Sometimes it only takes a slight change or enhancement to have a big effect.

In this section, you will learn about the following:

- The best methods for Up-selling or Cross-selling
- Using the Shopping Cart as an asset
- The Checkout process
- The general Payment Procedures used for purchases
- The best ways to prevent Fraud
- Keeping your Fulfillment/Shipping processes in order
- The importance of Inventory Management
- What to consider when Merchandising
- Using Feedback and User Ratings
- Providing Gift Services
- B2B Operations
- Processes for Soft Goods
- The benefits of Gift Cards
- How to leverage your Customer Loyalty Programs
- How Wish Lists can lead to sales growth
- The importance of Order Management

Part II: Selling

eCommerce Best Practices

2.1 Up-selling and Cross-selling

2.1.1   Introduction
2.1.2   What is Up-Selling?
2.1.3   Up Selling Hints
2.1.4   What is Cross-Selling?
2.1.5   Placement for Effective Selling
2.1.6   Up-Sell and Cross-Sell Success - Know Your Customer

2.1.1   Introduction

When customers go online to shop they are either researching products that they have an interest in or they are intending on making a purchase during that session. Whether it is just browsing through the site or selecting items off your site, you want to make sure that there were no missed opportunities that could increase the bottom line. Up-selling and cross-selling are two proactive techniques that encourage customers to add additional items to the cart.

Up-selling and cross-selling not only increases the total dollar amount of each transaction, but it enhances the customer experience as well. These tools can be customized to infer the types of products that would interest the shopper based on the products he or she is currently viewing or it can even go as deep as analyzing the historical shopping behavior and recommending items based on a longer span of collective user profile data.

2.1.2   What is Up-selling?

Up-selling is offering an upgrade, so to speak, to the product the customer is currently viewing or has already selected. Think of it like super sizing your meal—instead of buying a 1 GB thumbdrive, what about a 4 GB thumbdrive? Generally, the purpose of up-selling is to increase the dollar amount in the shopping cart.

> **All the suggested items should be related to the product and should not be presented in a way that takes up the customers' time.**

Though customers may not always take you up on your offer, they typically welcome these product recommendations because it is a sign of a personalized user experience. It shows them that you are interested in their interests. Unlike offline brick and mortar stores where customers are more suspect of

eCommerce Best Practices

employees' intentions of up-selling—whether it is based on commission—online up-selling generally does not stir up those weary feelings.

2.1.3   Up-selling Hints

Some hasty retailers opt to rely upon traditional customer mindsets—mirroring online tactics with those used at the offline counters of the retail store. However, there are better tactics to more effectively up sell online.

Customer's Mindset is Important

Since customers are already in the buying process, this is your chance to capitalize on their buying enthusiasm. The first thing you want to happen when up-selling is to get the user thinking about upgrading the product. You should be mindful of the placement and pricing of recommended items because the last thing you want to do is prevent them from making any purchase. In an eCommerce world you have assume every customer is willing to add or change items to their shopping cart, whereas in an offline store employees can judge which customers are more susceptible to up-sells.

You also need to make sure the products recommended are of the same nature of the products the customer has selected or viewed. Recommending an unrelated product requires the customer to make a greater leap in terms of changing their order. It is much easier for them to process an up-sell when it is related to the product they intended on buying.

Up-sell Needs Placement and Positioning

Within the retail site there are many areas where you can make up-sells, but they most typical pages to display them are: product display page, shopping cart pages, checkout page, and credit card entry page. It is best to up sell when the customer is more flexible their spending as opposed to feeling committed to buying an item, for example when they are making initial product selections or when they are reviewing the items placed in their cart. Last minute items appearing just before the submit button in the credit card page can also tempt the user.

Number of Up-sell Items is Vital

Some retailers believe the more items you offer the more chances the customer has to upgrade.

> Do not cross-sell or up-sell more than 3 items to the shopper for maximum returns.

Part II: Selling

www.ecommerce-best-practices.com

eCommerce Best Practices

However, this is not necessarily the case. Apart from having the right products to up sell, you also need to consider the number of products to up sell. Testing is required to determine the optimal amount of products to recommend, however the standard practice is not to entertain the shopper with more than three products.

Time Restraints on Sale Prices

Some retailers set time restraints on up sell offers, for example 'for a limited time only get this product for X amount of dollars.' This gives customers a little more incentive for changing their shopping cart because they feel like they might lose out on the offer if they do not upgrade.

> **Setting time restraints on up-sell item offers will nudge the customer to upgrade their shopping cart items.**

Be Different Every Time

Analyze the up-sell strategies that worked versus those that were not as successful. Once you have a sound system in place, adapt the up sell schema to reflect seasonality, price fluctuation and inventory. Testing and analysis is important here because the more trial and error you have in the short-term will contribute to a long-term formula to success.

The above suggestions are some easy and cost-effective best practices in making your up sell process effective and lucrative over time.

2.1.4 What is Cross-selling?

The purpose of cross-selling is to increase the total number of items in the shopping cart. Whereas up-selling suggests a 'do you want to super size that?' mentality, cross-selling implies a 'do you want fries with that?' school of thought. Cross-selling recommendations are related items that compliment the product the customer is going to purchase.

Cross selling items include competing, and accessory products. Competing products are alternative brand names with variations of prices and features. Using digital cameras as an example, competing products would be offering a Canon or Kodak camera when the product of interest is a Nikon. Complimentary products are those related to the product of interest, like photo printer. Accessory products are add-ons like a camera case or tripod. You can

use a combination of cross-selling types on one display window, but avoid overwhelming the customer with too many multi-variant options.

2.1.5   Placement for Effective Selling

The risk of cross-selling or up-selling is that it may make your customer question the item he or she thought about purchasing initially. To minimize this, placement of your recommendations is extremely important. The key is to avoid up or cross-selling once the customer has decided what items he or she will purchase. The pages you should focus your efforts on are:

Product Page

The product description page is a prime location for up and cross-sells because the customer is in an influential state. He or she has not committed to making a purchase yet and is open to exploring his or her options. This page is particularly advantageous because if the customer is reading through the description and ends up not liking the item, you have alternative options with the same interest.

Shopping Cart Page

When customers view the shopping cart page, as opposed to directly going to checkout, it is generally to review the items placed in the cart. You can infer that the customer is evaluating the purchase or checking to see if they selected the correct options. This presents another opportunity to up or cross-sell. All pertinent information should be displayed before the fold—products selected, price, size, color, promotions, as well as the up-sells and cross-sells. The recommendations should be apparent on the shopping cart page without having to scroll down.

Give Weight to Your Up and Cross-sells by Ownership

Taking up-selling and cross-selling a bit deeper, you can enhance their effectiveness by influencing customers beyond the product listing. If you apply ownership to the type of people recommending the up or cross-sells, it will give customers a better frame of reference for ordering the product. For example, listing the products that critics or professionals use will give more weight and incentive for customers to edit their shopping cart.

Customers are influenced by their peers and people with authority, and classifying your up and cross-sells will encourage enhancing their order. The key is to make the sells more relatable to the customer. Not placing ownership to the up or cross-sell translates to a less personalized experience than they could have.

2.1.6 Up-sell and Cross-sell Success—Know Your Customer

In order to know what products to up or cross-sell, you first need to know your customer. Learning about your customer starts with registration. The initial steps of creating customer accounts go a long way in creating a longstanding user profile. This data is invaluable not only for making up and cross sales but also for long-term growth with that customer.

As the customer shops from your site, or even just browses for that matter, there are Commerce tools that are able to track that individual's behavior and build a profile based on his or her click thrus. This data fuels the type products that will ultimately be suggested for up and cross-sells. The initial product the customer is interested in is enough to make a suggestion, but not enough to make the suggestions personalized.

But like everything else in eCommerce, you have to test what works best for you and for your customers. Find out which placement page works the best, which layout is optimal, whose suggestions are most influential, etc. The only way to know for sure what up-selling or cross-selling method is most successful you need to test it.

eCommerce Best Practices

2.2 Shopping Cart

2.2.1 Introduction
2.2.2 Abandonment of Carts
2.2.3 Shopping Cart Functionalities
2.2.4 Multiple Shopping Cars
2.2.5 What You Need - Ideal Shopping Cart

2.2.1 Introduction

The shopping cart is the iconic symbol for eCommerce, and is one of the most important features of an eCommerce site, not only for retailers but for customers as well. This one feature is rich with information. It is the first area products go when they are selected by customers and the last area products are held before going to checkout. The cart answers what the customer is interested in be it brands, types of products, or price range. As the visits to your site increase, the more data is collected to shape each user.

Another hot topic about shopping carts besides discovering the customer's interests is shopping cart abandonment. A customer can leave their cart in the middle of their shopping experience for many reasons, and it is up to the retailer to analyze why it happened and how it can be reduced. To do this, you need to know when and where each customer is likely to abandon their shopping cart.

> **The shopping cart must be visible to the user at all times. The cart icon should be available at several locations on one page.**

2.2.2 Shopping Cart Abandonment

There are many reasons why a shopper abandons their shopping cart during the middle or end of their shopping process. One of the most common reasons is shipping costs. A lot of shoppers select items without having an idea of how much their shipping costs will be. The customer has an idea of how much the purchase order will cost, and are often jolted after they see the total price after shipping and handling is added on.

Shoppers also abandon their cart if they do not like how the cart is designed, whether it feels comfortable and secure, or if there are too many processes

Part II: Selling

involved to get the grand total. Many retailers are opting for an AJAX based shopping cart that not only allows users to add items to the cart, but view all items and shipping costs in a sidebar on the page. As items are added to the cart, you will be able to see the addition in the sidebar and the new total order price.

Retailers aim to build a perfect shopping cart that has all the elements of what customers want to see and experience. To reduce the abandonment process we need to improve the shopping process. Retailers and marketers should work in tandem to streamline the process in order to attain the maximum ROI. Some of the options are:

- Simplify the steps of the shopping process until checkout. Visual references, like breadcrumbs, that mark the current step are also helpful for customers. Customers should not be distracted when making purchases.

- Customers often shop casually, place items into their cart and return later to complete the sale. Not being able to save this information for return visitors negatively impacts the user experience, limits their own ability to complete latent sales and loses critical customer tracking and behavioral data. In addition, returning visitors should receive clear indicators that you recognize them and know that they have items waiting in their shopping cart.

- Provide help either through customer care support numbers or online chat with your service department. Customer service should be available 24x7 and should be easily understood by customers visiting the site.

> **The cart must be easy to use with the standard conventional buttons even if the site is built on new technologies.**

- Back buttons are helpful as sometimes customers need to go back and rethink their actions without having to re-search the product altogether.

- Elements such as stock availability, editable products, wish list addition options, and email cart to friends are some best practice methods to ensure the customer does not abandon his or her cart.

- Email those customers that have products remaining in their cart. Retailers need to find unique methods to make them remember and

eCommerce Best Practices

encourage their purchases. Time is a critical factor and in such cases you need to act fast.

One of the best methods to assess cart abandonment and plan your strategy is to look at the data. Monitor customer paths throughout the process and look at the types of products abandoned. Path analysis is important because it shows where the shopper went prior to completing their checkout—whether it was a new product that grabbed their attention in another section or a third party banner.

2.2.3 Shopping Cart Functionalities

When viewing the shopping cart a customer should be given the following functionalities:

- Add Item to Cart
- Ability to update item quantities,
- Ability to remove items from cart,
- Display subtotals,

Some optional functionalities that a customer might benefit with are:

- Shipping Cost,
- Tax Total,
- Cross-sells and Up-Sells,
- Option to buy an item later,
- Option to add the item to a wish list,
- Inventory Check,
- Price Check,
- Merge Orders

The inventory and price check functionalities would come into play if the user had previously saved the cart and is returning to it after some time. During that time, the item in the cart may no longer be available or its price might have changed. It is important to notify the user of this change when the user views the cart page to prevent any confusion and miscommunication.

The Merge Order functionality is useful when a customer comes to an eCommerce site, places items into the shopping cart and then logs in. This might lead to the case where the customer's saved cart might already contain items. The merge functionality would merge the items in the two carts so that the customer might not lose any of the items that they wish to purchase (both in their anonymous cart and their saved cart). It becomes a business decision at that point, if the one cart should overwrite the other or the items should actually be merged.

Other functionalities such as the ability to enter shipping information, billing information, apply promotions etc. are covered in the Checkout section of this book.

2.2.4 Multiple Shopping Carts

If applicable your site might also want to offer the user the ability to have multiple shopping carts. The cases where you might want to do that is when there is an approval process that might be associated with the cart (either internal to the customer site or within your organization). The order approval process is usually associated with a B2B site. The ability to have multiple shopping carts would allow the customer to associate different costs with different budgets (where applicable) and essentially split the cart where there is a chance that only some items might be approved and others rejected for purchase.

Another reason for offering multiple shopping carts would be if your site offers configurable products. The scenario might arise where a user is configuring multiple products and is only ready to place an order for one of them. By splitting the orders (into multiple shopping carts), the user can place one portion of the order and save the remaining configuration for a more convenient time.

With multiple shopping carts, the complexity of the eCommerce implementation also increases. Users must be offered an intuitive way to switch between shopping carts, care must be taken that items that are added to a cart goes into the right cart, ability to move items between carts and ofcourse non-ambiguous management and display of the carts themselves.

eCommerce Best Practices

2.2.5 The Ideal Shopping Cart

Building the right shopping cart is not a perfect science, and nothing is absolutely foolproof. However, there are some tips that could enable you to assist all your customers towards the checkout page and complete their orders.

Cart from the Outside
The cart placement and its responsiveness add to a positive customer experience and retailers should take special notice to see that the below points are covered. These are applicable when the customer is going through the shopping process of adding more items to their cart.

> Placement of the cart and what's inside hold equal value. Marketers should make sure the information in the cart is helpful and easy to understand.

Cart as an Icon

The cart is the focal point of the shopper's experience. Customers should be on the path to purchase, and if the cart is visible at all times and is conveniently placed within the navigation framework, then there is little hindrance in the shopping process. Online shoppers need to know both the number of items in their cart and also the total cost of their purchase. The cart should remind and inform the shopper of these concerns.

Action of Addition to be Visible

Whenever the shopper adds a new item to their cart, sometimes they are not aware that this action has been performed. With the advent of new technologies like AJAX and Flash, the customer can remain on the same page while only the numbers and cost below the cart icon get updated. So how do you make the customer aware of this action without having to leave their product page? Retailers should build carts that are more functional in presence and provide the customer a signal that notifies them that their command has been performed successfully.

Cart from the Inside
Apart from the basic tips to be affective in the look and feel of your shopping cart, equal importance should be given to what and how you present the tools within your cart. Some of the effective methods which can be followed are:

Always Disclose Costs

Apart from the display of costs during the shopping process on the cart icon, you also need to display in detail the costs of the items, shipping costs, and tax if applicable. Free offers on shipping for purchases that reach a certain total price should also be prominent here to entice the user to proceed with their check-out or add more items to the cart in order to get the free shipping.

User Friendly Tools

Helpful tools like "save for later", "wish list", "email a friend" and "print" are some ways you can prevent the customer from abandoning their carts due to the total high price or some other reason. Standard tools like edit, delete/remove, update quantity, and add new items in the cart are some ways to give your customer control over their cart and shopping process. Sometimes retailers also give the option to change some of the styles of their products without leaving the cart, for example selecting a new color for a sweater or size for pants.

Presence of Loyalty Programs

Display clearly all loyalty programs as it is a powerful tool to build customer relationships. It adds a reason to prevent the customer from not completing the checkout process. If customers are surprised at seeing a high price tag on their purchase and know they are likely to revisit the site again, loyalty programs can nudge the customer towards completing his or her purchase because they know it will contribute to future benefits or discounts.

Customer Help

Contextual help is another important factor that improves shopping cart usability. Providing your customers with answers without affecting the navigation flow increases order size and reduces cart abandonment. Prominently displayed links to standard shipping charges, FAQs, and return policies should be available in order to readily provide the information customers commonly seek out.

eCommerce Best Practices

2.3 Checkout

2.3.1 Introduction
2.3.2 Checkout Flow
2.3.3 Checkout Points to Consider

2.3.1 Introduction

The checkout process is as important as the shopping cart, if not more so, because any complications with checkout may deter customers from completing that order or any future order from your site. If the number of pages to checkout is too little or too much it can make or break sales. Retailers must make sure they leverage all their strategies that lead the customer through the checkout process.

The customer should be aware of what step in the checkout process he or she is in, and how many steps are left to complete the purchase order. In some eCommerce sites, customers have the option of saving their billing and shipping information to their account, which allows them to have a 1-step checkout instead of going through the entire process each time they make a purchase. However, not every customer is comfortable with this, so you want to be sure that your checkout process makes every customer feel secure as well.

Besides having a sound checkout process in place, it is important to encourage the customer to actually checkout. Making action buttons like "proceed to checkout" clearly visible on product pages and at the top of each page let the customer know he or she can use this function at anytime to complete the order. The last thing the customer should have to do is search for the checkout button. Many eCommerce sites experience missed opportunities because of poor design and layout.

> **Give customer control of their orders—multiple shipping or billing addresses, ability to email their order confirmations.**

2.3.2 Checkout Flow

The checkout flow refers to the steps that the user takes when they have the necessary items in the cart and are ready to purchase them. The main pieces of information that need to be captured are shipping and billing information. The rest of this section lists out the elements a user might encounter in a checkout flow. However not all the elements mentioned below may fit into your eCommerce site and there is certainly flexibility to how the information is captured (multiple pages or single page).

The standard checkout flow consists of starting from the cart page. This is where the customer can edit item quantities, remove items, enter promotional codes and view the sub-totals which may include shipping and tax amounts. It might also show some promotional items to encourage the user to add more to their shopping cart such as informing them that they would receive free shipping if they purchase merchandise over a certain amount. The user might be given the ability to calculate the shipping and tax amounts on the page itself.

The next page in the checkout flow is typically the shipping information page. Here a user can select a shipping address that they had previously saved in their account or enter a new address. It is on this page that an order might be split into multiple shipping groups or special instructions entered for the order.

The billing information page follows the shipping information page. The user either uses an existing payment method or chooses to enter a new one. If entering a new payment method, the user is typically given the option of using their shipping address for the billing address. The payment methods can be credit card, gift card, store credit, online check or bill me later type option. Each of these would require different sets of information to proceed. Any promotional codes might also be entered into this page. If the site offers, a combination of the above payment methods might be entered by the user. On submission of the billing information page, the payment methods are typically validated either using an in-house check or via integration with a third party tool.

After submitting the billing information, a confirmation page is shown which summarizes all the user's information and asks the user to confirm the order. At this point the order is submitted for fulfillment and payment methods authorized for the given amounts.

The above is just one way the check out flow might be organized. Alternatives that you might consider for your organization might be:

- Combining all three cart, shipping and billing pages,
- Combining shipping and billing pages,
- Switching the order in which shipping and billing information is captured
- Not showing confirmation page

eCommerce Best Practices

Some user would prefer to have one company handle all their checkout information for security purposes. To capture the orders from that customer segment, your site might opt to use a third-party checkout process. A couple of popular checkout processes Paypal Checkout and Google Checkout.

2.3.3 Checkout Points to Consider

Once shoppers have added all their items in their shopping cart, they should not have to constantly confirm their order in the checkout phase. The shopping cart should be the page where they assess their order and the checkout is where they complete their order. So how do you prevent customers from abandoning their order? Though there are no standard practices that guarantee success, it usually begins with a trial and run process, along with the assistance of analytics to steer you in the right direction. Some points to consider are:

Security

Once in the checkout page, you should reassure the shopper with a globally known third party security firm, for example, ScanAlert or Verisign. This increases stability and shows that your brand is associated with quality and recommended guarantees. Displaying

> **Showing third-party vendors are important, as it increases consumer confidence**

information banners like 'satisfaction guaranteed or your money back,' may seem like a good tactic, but be careful as it might actually backfire and make them question your reliability. If your user interface and experience is enjoyable, the last thing you want to do is make them question your security.

Another added level of security can be introduced is to require the user to enter their password on checkout if they haven't done so in the current session. This conveys to the customer that no unauthorized purchases can be made if they are auto-logged in and ensures no one misuses an account which has auto-login enabled. However asking for the password introduces an extra step in the checkout process which may lead to more abandoned carts.

Giving the Customers What They Want

Smart links with pop-ups that incorporate critical information build a strong

eCommerce Best Practices

customer service culture. Clear messaging around free shipping or how a shopper can get discounts on a particular order are some subjects that should be highlighted as an example. Sometimes offering repurposed billing or ship to opt-in boxes that automatically populate previously entered fields is a smart shopping tactic. Loyalty Programs coupon codes or gift vouchers should be presented carefully as you do not want the consumer to abandon their shopping experience by having to search the slot for entering a coupon code they already have or look for when adding a gift to their cart. Strategically placement of these can entice and compel your customer to complete their checkout with ease and no confusion. Promoting and providing an easy mechanism to add gift cards to an order is yet another small, but effective element that can add to the total size of your shopping cart.

> **Promote Gift cards and coupons with clever placement before customers confirm order.**

Give them Options

As mentioned earlier, when customers are in the shopping cart and subsequently in their checkout process, give them the option to save, email or add multiple addresses to their checkout page. This gives customers control and increases the probability that they will return to your site should they abandon it.

User-friendly Icons

Even during the checkout process, the steps should be minimal to reduce the likelihood that the customer experiences a moment of frustration or confusion. You do not want to push the customer out of the checkout process sooner than they had planned. All interactive icons should

> **Allow new users to complete checkout without registering. Email address procured at end provides strategy to target them later for registering options bundled with offers.**

be standard and conventional so that even if the site is in rich media, the technology would not deter the user from completing the checkout process.

Welcome New Users

The best way to ensure that all your customers, whether new or old, complete a checkout process, is to avoid forcing registration. Casual visitors and new users should get the option to complete their checkout without having to register. If at all possible, one could start by just having their email ids as a registration option so that incase they cannot complete the checkout, you could target them later with shopping incentives based on their abandoned carts.

Shipping/Tax Details

Customers seem to shy away when the shipping details are not mentioned apart from the price. You could also include charts which tell the customer about the approximate days it would take to ship through a variety of vendors. Tax is also crucial and retailers can give customers option to select which zone they want to know about the tax.

Promotions

Keep advertising special promotions or coupons up to the billing information page. Make sure the promotional code entry box is visible on the page before the fold. Promotions are highly effective tools for completing orders, free shipping promotions in particular.

eCommerce Best Practices

2.4 Payment Processing

2.4.1 Introduction
2.4.2 Terms and Definitions
2.4.3 Forms of Payment
2.4.4 Miscellaneous Methods and the Future

2.4.1 Introduction

It is an eCommerce site's sole function is to display a list of products (wares) and entice consumers to make a purchase. The site should be able to help them search and find the product, or products, they want or need. Once the consumers choose the product, it is the site's job to process the orders from the consumer; part of which will include the payment information. Any basic eCommerce site must possess the ability to process payment information, be it taking the credit card information in clear text emails like in the mid 1990s or storing it in a sophisticated encryption form as a part of the order. Currently, the forms of payment on an eCommerce site can vary from the most common form of legal tender to the more intuitive forms that may involve credit cards, gift cards, or even store-reward points.

The execution and implementation of the payment procedure for eCommerce sites is multi-faceted. Different business problems and some misconceptions in payment processing, as well as the origin of different payment methods and how they are processed in an online scenario should also be known.

> **Involve the Corporate Data Security Team Early in the project to comply with the standards. Plan on periodic penetration tests to identify security flaws.**

First you need to know the different terms used in an online transaction based environment. There are traditional payment methods and some unconventional payment methods, where the more traditional payment methods for an online transaction consist of credit cards, gift cards and third-party payment procedures. The unconventional payment methods for an online transaction are cash and/or check.

2.4.2 Terms and Definitions

In an online world, it is always easy to understand concepts and different procedures once you understand the terminology and definitions. In payment processing, there are certain common definitions and terms that you should know which will make understanding the procedures a lot less confusing.

Payment Gateway / Gateway

A "payment gateway" or simply a "gateway" facilitates the online payment (payment transaction) between the merchant and the acquiring bank. The bank, once it has processed the transaction, will forward the acceptance or rejection of the gateway to be sent back to the origin of transaction. The coining of the term gateway comes from Visa cards' definition. In the MasterCard world we encounter something called Data Storage Entity (DSE). The reason it is called DSE is because when a Gateway receives information, it typically stores some of the data before transmitting it to the acquirer's bank.

Merchant Provider (Independent Sales Organizations)

These are the organizations that maintain the merchant account information and are responsible for acquiring the funds from the customer's bank to the merchant's bank. Nowadays a lot of the gateways and ISOs are performing these jobs as a single entity.

Cardholder Information Security Program (CISP)

A Visa program that establishes data security standards, procedures, and tools for all entities—merchants, service providers, issuers, and merchant banks—that store Visa cardholder account information. CISP compliance is mandatory and a compliance test is done before launching any eCommerce site.

> **Build a check list of the compliance tests that need to be complete for a launch and work towards meeting them (CISP).**

Address Verification Service (AVS)

AVS is a process where a merchant, when involved in a card not present transaction (over the phone, online/eCommerce transactions), sends and compares the billing address provided by the customer with the address provided by the customer to the issuing bank. This process helps merchants reduce the risk of fraudulent transactions.

eCommerce Best Practices

Authorization

Authorization is the process by which a card issuer approves or declines a Visa card purchase. This process involves reserving funds from the customer's bank for the goods purchased or services rendered.

CVV2 / CVC2

Card verification value 2 (CVV2) or card verification code 2 (CVC2) is a three digit code printed on the back of the card to ensure that the credit card is in the hands of the customer. CVV2 is a term used in the Visa world. CVC2 is a term used in MasterCard's world. This information is passed in the transaction for authorizing the card. Merchants are prohibited from storing this value.

Chargeback

A transaction is returned to the merchant bank by the customer's bank, usually due to a disputed transaction. The merchant bank may then resolve the transaction or "chargeback" the transaction to the merchant. A merchant is considered a high-risk merchant if there are high-risk (chargeback) transactions due to the nature of the business or because of poor business practices. Visa has a program called Merchant Chargeback Monitoring Program (MCMP) to check if a merchant has had more than one percent chargeback. The merchants are then required to work with the banks to reduce the charge backs to avoid fines or termination of the business agreement.

> **Implement a fraud prevention mechanism to reduce chargebacks and be identified as a high risk customer.**

Due to the complexities in processing charge backs mostly because the credit/charge card companies often change the rules, there has been a wide opportunity for a whole new set of businesses that provide services for chargeback handling.

Interchange and Interchange Fee

The process of authorization and settlement of credit card transactions through the gateways is called Interchange. The amount the card associations charge the acquirers for each

eCommerce Best Practices

transaction is called the Interchange fee. These fees are expenses incurred by the card issuing bank for providing the credit to the cardholders. The fees are passed on to the merchant by the acquirers.

2.4.3   Forms of Payment

There are a variety of conventional payment methods that are processed in the online world. The consumers have the option to use credit cards, gift cards, pay by check, etc. to make payments.

> **Spend time upfront developing a strategy that allows you to test and get the kinks worked out of the different payment methods.**

The basic process for each payment method is the same and the path taken by the transactions are similar. To understand the processing, each payment method will be described with the aid of the figure below. This image of processing is valid for all three forms of payment.

eCommerce Best Practices

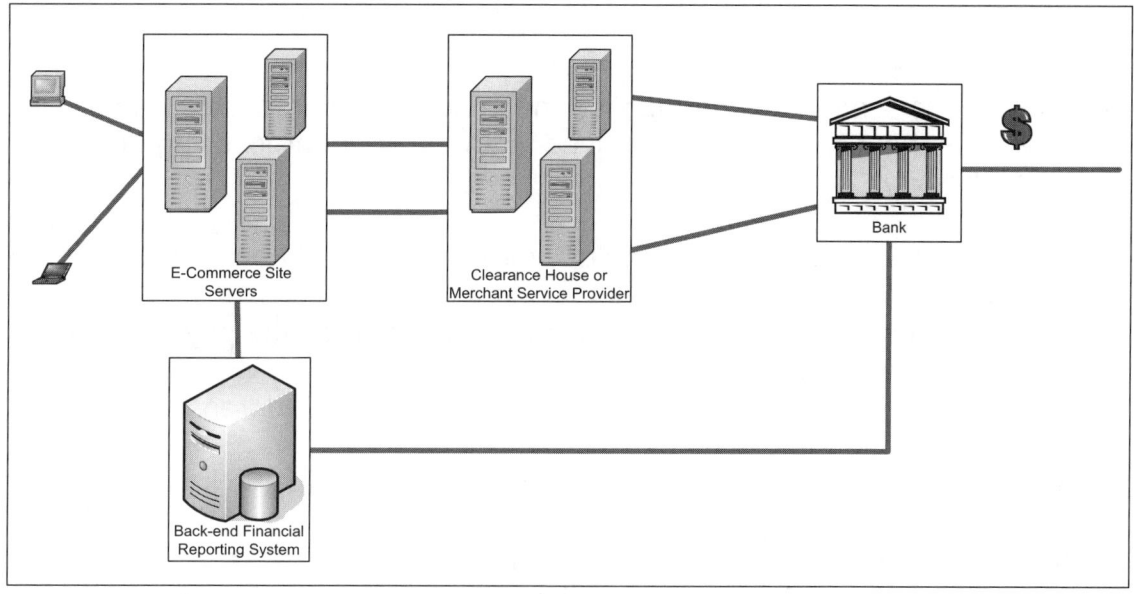

Figure: This figure shows the important components that are involved in a typical eCommerce Website.

Credit/Charge Cards

Credit cards have been around for a long time. They are different from charge cards, although nowadays the term charge card and credit card are used interchangeably. The basic difference between these two is that a charge card requires the balance be paid in full every month and in a credit card a balance can be maintained at a cost of having interest charged to that balance. You will learn about the payment in the normal "Brick and Mortar Store" Scenario and how the same sort of transaction happens on-line.

Physical Retail Store Scenario:

When the customer acquires goods or services and pays with a credit card, he or she hands the card to the merchant, to which the card is swiped on a credit card terminal to initiate a transaction. At this point the information is sent to a credit card processing network and is authorized (funds reserved) for the price of the goods or services rendered. This scenario has the customer physically possessing the card; these transactions are called customer present transactions. The incidences of fraud in this scenario are very little, as the

identification of the card holder can be verified before the transaction. This in turn leads to fewer chargebacks.

eCommerce Store:
To explain this scenario, take a look at the image in the preceding page. The following are the different steps involved in the payment processing of an eCommerce site.

- The customer comes to the eCommerce site and having selected the goods, proceeds to place an online order.
- The customer enters the payment information on the payment page and submits the order.
- The credit card information along with the billing address and CVV2 are sent to the Gateway.
- The Gateway does minimal verification and sends the information to the merchant provider.
- The merchant provider will send this information to the acquiring bank.
- The acquiring bank will send the information back to the customer's bank for authorization.
- The customer's bank, upon receiving the authorization request, sends the response code back to the payment Gateway. The response code is also used to determine the reason for a failed transaction (such as insufficient funds or bank link down)
- This response code is sent back to the website to be interpreted and presented to the user.

At the end of the day most eCommerce sites send the authorization and order information to the back end financial processing systems. These systems, in turn, generate information for the acquiring bank to extract funds and deposit them into the merchant's account.

There are different ways as to how each piece of communication is processed. The information from the website to the Gateway is sent through SSL communication or the merchant can sometimes have an in-house Gateway which communicates with the acquiring bank.

The communication lines between the Gateway/merchant provider and the acquiring bank are usually leased lines.

eCommerce Best Practices

Role of AVS and CVV2/CVC2:
When the credit card authorization transaction is sent across, part of the information contains the billing address provided by the customer. When the information is received at the customer's bank the billing address is matched with the address in the registry of that customers account. If this is a match then a positive AVS code response is sent back to the eCommerce site. Similar to AVS the transaction also contains CVV2/CVC2 matching. These mechanisms are necessary for reducing the incidences of fraud.

It is also in good practice to work with a bank in getting a periodic data feed of a list of compromised credit cards. This check will also protect you from chargebacks.

Gift Card

Scrip, essentially substituted money, is seen as the precursor for the modern-day Gift card and Gift Certificate. Gift cards are issued by the merchant. There is no way of identifying the ownership of a Gift card and because of this, if the card is stolen or lost, it is gone and there usually are no refunds. Gift cards are a great way for the companies to promote purchases in their stores; hence they are sometimes called a "Closed Loop". The business also earns interest on the float until the card is used.

> **When building the Gift Card processing system, keep it close to the credit card processing methodology of authorize and settle.**

Gift cards do not have value on them until a value is added to them. The value of the card is typically not stored on the card. It is maintained by the merchant or by third party processing companies. One disadvantage is that some of the gift cards have a maintenance fee after a period of time and is deducted from the card, there by reducing the value of the card. Some states like Massachusetts have laws that eliminate the non-use or expiration date of the gift card.

There are two approaches that companies take for all of the gift card processing. The first approach is where everything is processed within the merchant's organization. In the second scenario both the issuing and settlement of the gift card is done by an outside resource like the Stored Value Systems (SVS)

www.ecommerce-best-practices.com   83   Part II: Selling

eCommerce Best Practices

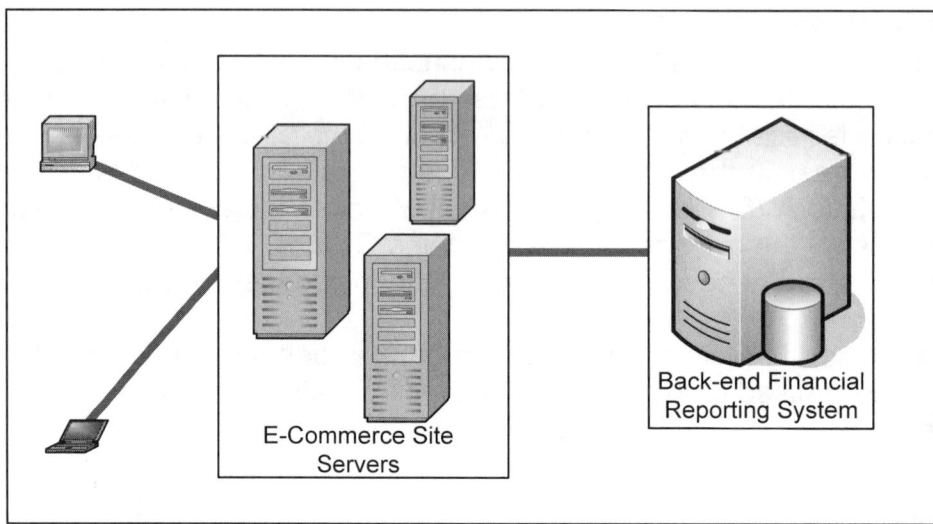

In the first scenario when a Gift card is issued, the card number, pin and value are maintained in the merchants issuing system. Once the order is placed, the entire processing takes place with in the merchant's realm where the card is verified and payment is applied against the card. See figure above.

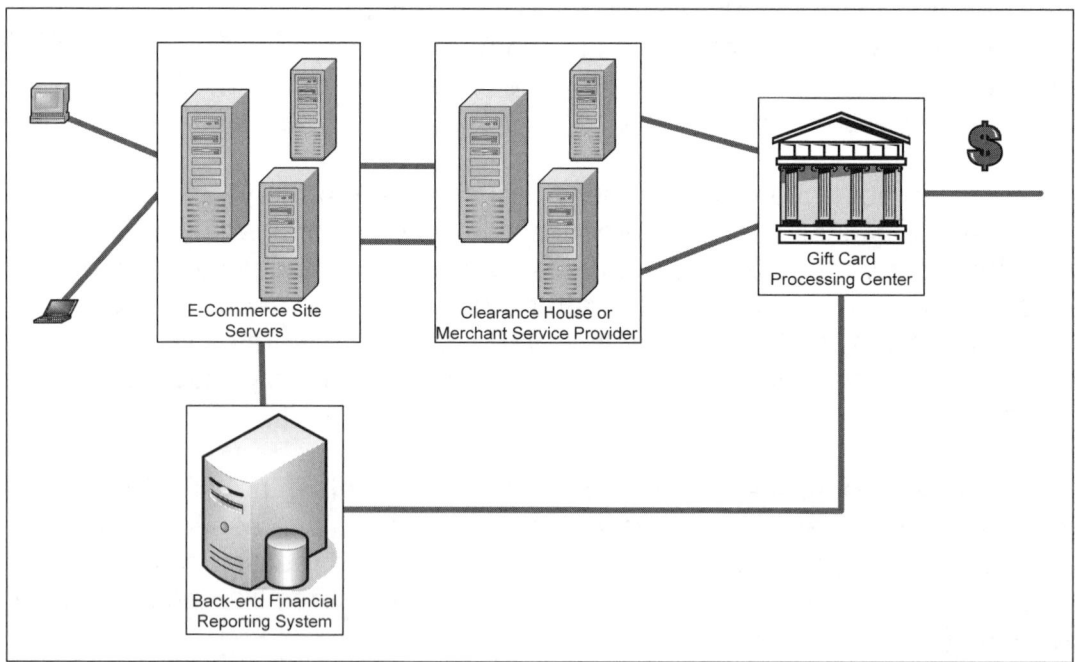

In the second scenario, once the customer buys a card the cards are valued by a transaction similar to the above processing of the credit cards. Once it is

eCommerce Best Practices

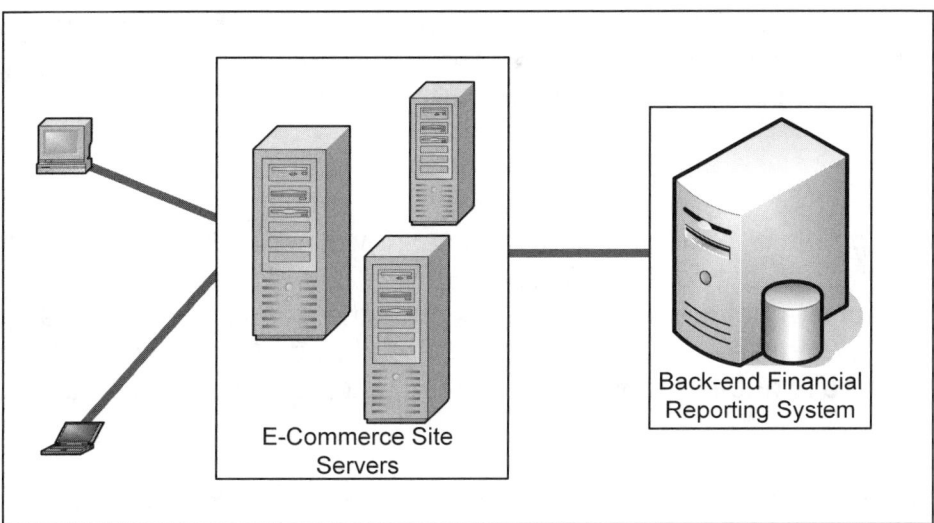

In the first scenario when a Gift card is issued, the card number, pin and value are maintained in the merchants issuing system. Once the order is placed, the entire processing takes place with in the merchant's realm where the card is verified and payment is applied against the card. See figure above.

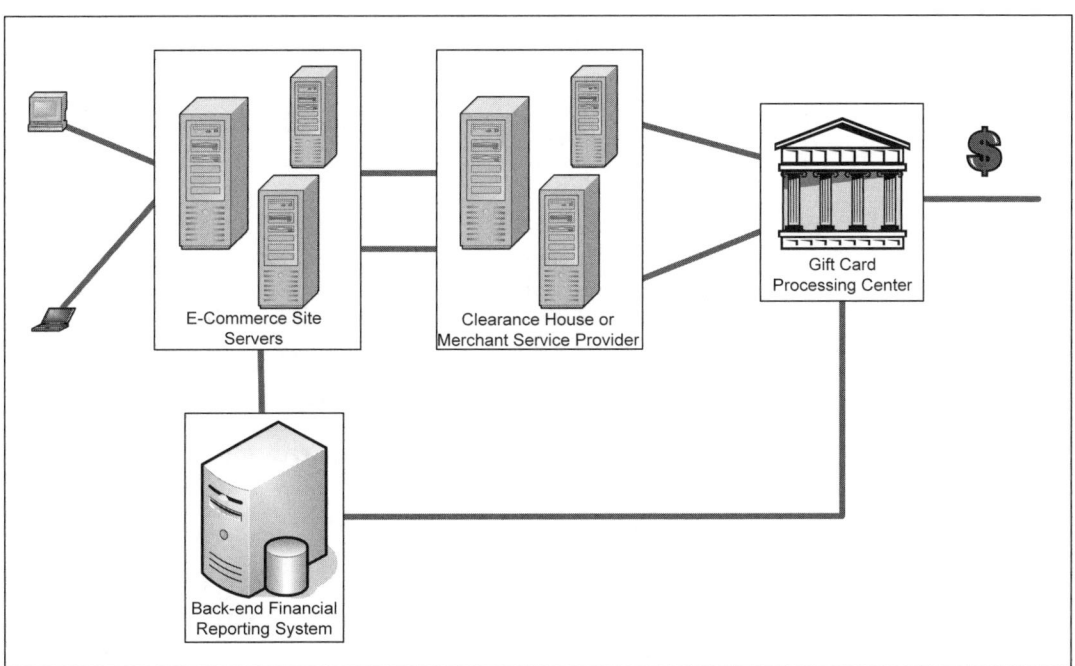

In the second scenario, once the customer buys a card the cards are valued by a transaction similar to the above processing of the credit cards. Once it is

eCommerce Best Practices

order. Paypal and other services like Paypal have also been providing API that will enable you to make calls to their servers to settle payments, thus, providing another approach to make purchases with.

Cash

The main problem for sites that accept cash payments through their retail stores or through services like Western Union is that the merchant, traditionally, can only settle a payment process only once the goods have been shipped from the warehouse or if the services are rendered. With the cash payment method there is nothing to settle once the customer selects this payment method.

The businesses are getting around that by making the consumer aware of how the funds would be returned if there is a problem in the warehouse, etc.

eCommerce Best Practices

2.5 Fraud Prevention

2.5.1 Introduction
2.5.2 Fraud Prevention Techniques
2.5.3 Fraud Prevention Framework

2.5.1 Introduction

There is no arguing that the ability to prevent fraud has become a survival issue for eBusiness operations. It is estimated that $3 billion1 was the revenue lost to fraud from eCommerce sites in 2006. Fraudsters know this and plan to make future years even better for their team.

As an eBusiness operation, your goals should be to minimize chargebacks, reduce lost revenue in fraudulent orders, and avoid higher transaction rates and fees from banks. However, the panorama gets more complicated because you also want to keep your good customers and rightful revenue undisturbed, keep the process to review suspicious orders manageable and to keep your prevention systems ahead of fraudsters.

Simply speaking, fraud prevention is the strategy you need to implement to avoid or mitigate the occurrence of fraud. For any business, traditional or eBusiness alike, that strategy can be as comprehensive as the business itself since every business operation is susceptible to fraud. One area that is of common interest to all eCommerce businesses and the one we will discuss in this chapter is the prevention of fraud on credit card transactions.

Credit cards are commonly the first payment method accepted by eCommerce sites and the one that most online customers use. Online transactions using credit cards are referred as card not present transactions (obviously because neither the physical card nor the consumer are present), just like when you place an order by mail or on the phone. This very fact makes it more attractive to fraudsters who see anonymity as an ally.

> **Because of the relative anonymity of online purchases, fraudsters seek to exploit this weakness.**

---

[1] According to Cyber Source's 8th Annual Online Fraud Report – 2007 Edition.

Starting from the basic elements and up, this chapter will discuss the most popular fraud prevention techniques. It is important to outline a framework in which these techniques are commonly used in an eCommerce site. There should be an emphasis on the need for constant change, the ability to model the fraud prevention strategy to your business needs, and the importance of the analysis of fraud. This is not a blueprint to win the battle against fraud in the card not present world, but it is a first step to get ready for the fight.

2.5.2   Fraud Prevention Techniques

In the eBusiness environment these days it is impossible to operate without covering the basics of fraud prevention. Your eCommerce site will be vulnerable to the attacks of everyone—from the experienced fraudster to those who are just testing how far they can go.

Fraud Prevention Techniques are the methods used to protect yourself and business from fraud. They are general, not specific in the way you can implement them. On the other hand, Fraud Prevention Tools are specific implementations of a technique by an organization or a vendor. By understanding the techniques we will be able to understand the different implementations and the way they operate in the context of a fraud prevention system.

At this point the notion of the fraud risk score gets introduced. Fraud Prevention is based on the identification or assessment of potential fraudulent transactions. This assessment defines the chances or possibilities of fraud, and when expressed as a number, it is called the fraud risk score. For now we will use this notion to establish examples of how the application of a technique contributes to this factor.

Negative and Positive Databases
Negative and Positive databases are sets of data elements like email, credit cards, address, loyalty number, country code, and others that are considered fraudulent or related to customers with problems in the past (Negative) or are considered a part of known customers in good standing (Positive).

These databases are the first lines of defense in the fraud prevention system since a match will either set a high fraud risk score (Negative Database) or accept the order as not fraudulent (Positive Database). Negative and Positive databases are used to take an immediate decision on the order. Some other databases, sometimes referred to as Suspected

eCommerce Best Practices

Fraud databases are used to contribute with the fraud risk score and will not make an immediate decision.

When an order matches a Negative Database the priority for a manual review is set to the lowest possible and asks the consumer to contact a Customer Service Representative.

Keeping these databases up to date is the key to making them effective. This includes, for example, the ability of promoting data automatically from a different sales channel into a negative database or the other way around by removing data if a customer answers the request for additional information and works with the Customer Service Representative to clear out the order.

Rule-based or Logic-based checks

Rule-based checks define a pattern for fraud suspicion by applying some logic that checks certain conditions in an order. Among the most popular checks we can find:

- Total order amount or ranges of total order amount.
- Orders with high risk products such as gift cards.
- Total number of products or total number of the same product in the order.
- Expedite Shipment (like overnight or 2nd day shipping).
- Different billing and shipping addresses.
- International orders.

By looking at each of the most popular rule-based checks, it is easy to understand that one of them alone is not enough indication of fraud, but as they are combined or contributed with the fraud risk score, the indication of fraud is more effective.

One of the key elements to make the rule-based checks effective is the ability to model them based on the analysis of the eCommerce site.

Finally, do not confuse rule-based checks with other non-fraud rules and restrictions implemented in the eCommerce site like shipping restrictions. They are different in nature, normally checked at different times and processed in a different way.

eCommerce Best Practices

Velocity Checks

Velocity checks look for data elements in the order that has been used in different transactions in a short period of time. Credit card number, billing or shipping address, email and phone number are among the data elements more frequently checked. Some additional elements that can be checked as part of the transaction are IP address and session parameters.

These checks are aimed to identify unusual behavior like a fraudster using multiple stolen credit cards and placing multiple orders, but using the same phone number or vice versa, with a fraudster trying to place as many orders as possible using one stolen credit card. Fraudsters tend to place multiple small orders to avoid some of the rule-based checks we mentioned before like total order amount and total number of products.

> Velocity checks aim to identify unusual transactions with credit cards by checking multiple data elements.

The number of occurrences and the period of time in which the velocity check is matched are key elements to make the velocity checks effective. These parameters are set based on the analysis of the behavior of a typical customer in your eCommerce site, for example, how many orders are placed per typical customer in a day or a week.

Address Verification System (AVS)

The Address Verification System is a service provided by the banks issuing credit cards and the credit card associations. It validates the billing address submitted with the credit card information during a transaction (normally during the credit card authorization) against the address that the bank keeps in its records.

As a normal practice resulting from integrating the eCommerce site and the payment processor, the AVS code will be a part of the authorization response. The AVS works by checking the billing address and returning a code that describes how the billing address matches the one kept by the bank.

The result of the verification can be expressed in the following terms: exact match, address matches and zip code does not match; zip code matches and address does not match; address and zip code do not match; address information unavailable, etc. The resulting code does not affect the result of the transactions. In other words, the authorization can be approved regardless of the resulting AVS code.

As mentioned with other fraud techniques, the AVS code alone is not an indication of fraud but an additional contributor to the fraud risk score. In a fraud scoring system, an AVS code indicating a full match combined with a rule-based check for the same billing and shipping address could lower the fraud risk score.

Suspected Fraud Databases

Similar to the Negative and Positive databases, these are sets of data elements like email, credit card, and others that contribute to the fraud risk score. Customers with a previous chargeback can be added here. Zip codes or even country codes with a high rate of fraud can be included.

Card Verification Method

This is based on the 3 or 4 numeric code on the back of the credit card for Visa and MasterCard, and on the front of the card for American Express.
The code is not available from the credit card magnetic strip and it is not printed in sales receipts where the credit card has been used. Additionally, credit card companies restrict retailers from storing this number in their systems or keeping it in their records after the transaction has been completed. Since the only way to provide the code is by looking at the card, it validates the possession of the card.

Additional Techniques

There are of course additional techniques, some of them in the early adoption phase like MasterCard SecureCode and Verify by Visa. Moreover, new techniques will be created as a result of the many challenges presented by

eCommerce Best Practices

fraudsters. At the end of the day, it is your ability to adapt to those challenges that will keep you ahead of the fraudsters.

2.5.3 Fraud Prevention Framework

Now that you have a basic understanding of different fraud prevention techniques it is time to put them together. Start outlining a simple fraud prevention framework that shows how an order flows between the different stages.

The anatomy of this simple fraud prevention framework consists of three (3) stages:

Validation: This stage includes validation techniques used before the order is placed. At this point, it is validated that the consumer has entered a valid credit card, valid expiration date, valid zip code and email address. The technique used to validate a credit card number is called mod 10 (for modulus 10). Mod 10 is a simple algorithm available as an out-of-the-box feature in most eCommerce applications.

Fraud Subsystem: This is where all the techniques mentioned in the previous sections are used. The outcome of this stage is either a valid order which is sent to the Order Management system or a potential fraudulent order (suspected fraud order) that is sent for manual review.

Manual Review: In this stage, suspected fraudulent orders are routed to staff members specially trained to deal with fraud. The fraud specialist will determine whether the order is in fact fraudulent or if it is a good order that needs to be processed. This is obviously an expensive task and as you optimize the fraud subsystem stage, you can minimize the number of orders for manual review.

The next diagram shows the three (3) stages of this simple fraud prevention framework:

eCommerce Best Practices

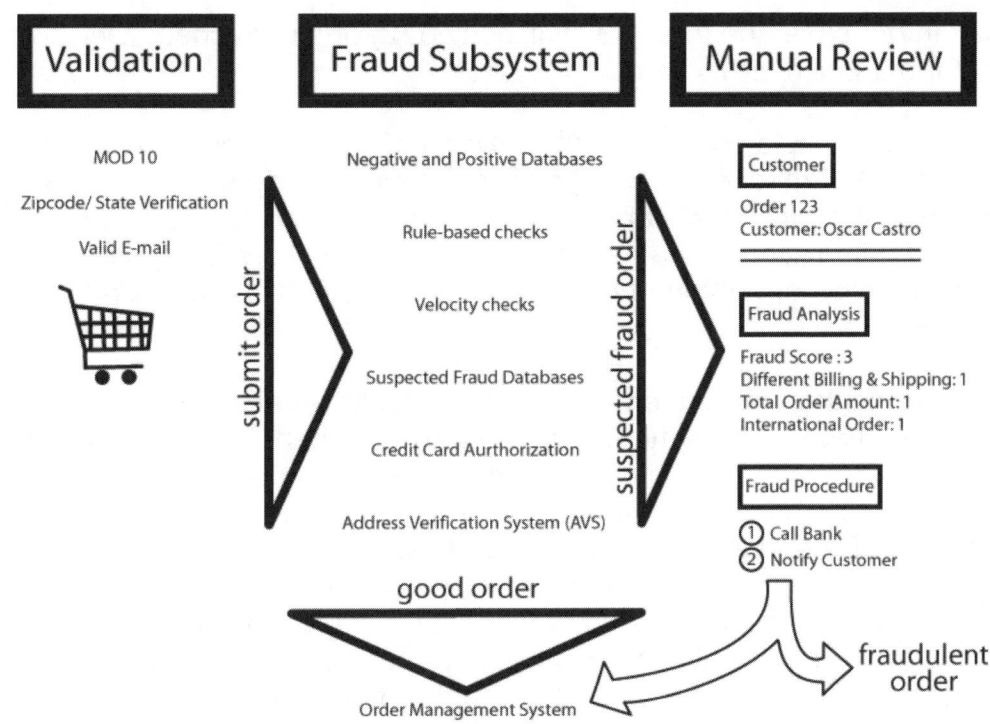

It is during the Fraud Subsystem stage that an order is identified as a suspected order of fraud. We mentioned before that when we assess the possibility of fraud we can represent this with a number that we call fraud risk score. Most of the fraud subsystems are implemented by using a scoring model in which every fraud technique provides a positive or negative score that is evaluated against a threshold.

As we can imagine, these models are complex and vary among different implementations. In essence, we need to analyze statistically the different characteristics of fraudulent orders and assign a score to the different fraud factors representing how strong their contribution is in the overall result. In a simple model, every factor has a fixed score; in more complex models, this score will depend on the different contributors being assessed.

After all the big words of models and statistics, try to model your own very simple scoring model. Every technique implemented in the fraud subsystem will have a score, if triggered.

When the contribution of that technique is strong, its score will be two (2); otherwise it will be one (1). The use and definition of a threshold also vary by

model. In our case, the threshold will be two (2) and if reached, it will send the order for manual review.

Based on our simple scoring model, several of the rule-based checks like Total order amount, Expedite shipment and International orders will score one (1) if triggered. A velocity check for credit card will score two (2). Just to make it a more interesting scenario, a negative database will score five (5) if triggered. Described below are different case scenarios and notice the different outcomes.

Case 1: Fraudster with stolen credit cards attacks an eCommerce site with multiple purchases. Very often, fraudsters use different billing and shipping addresses. From the beginning, if the fraudster attempts to place a high dollar amount order both the total order amount and the different billing and shipping techniques will be triggered, and the fraud risk factor will be two (2), sending the order for manual review. At this point, customer information could be promoted to the suspected fraud database. On the other hand, if the fraudster places low dollar amount orders, a velocity check could be triggered in subsequent orders for a final fraud risk factor of three (3) (including different billing and shipping addresses).

Case 2: Known fraudster tries to place an order. If the fraudster is on a negative database, the order is automatically sent for manual review with a fraud risk factor of five (5). Since the manual review task is expensive, you can decide to only route to your fraud staff orders sent for manual review with a fraud risk factor less than five (5). We would have also sent a notification to the customer requiring additional information and asking him or her to contact our call center.

Case 3: International Order. Evidently, if you need to review every international order, you need to set this rule-based check scoring two (2). However, by setting the score to one (1) you make sure that even different billing and shipping addresses will send the order for manual review. You could even use the Address Verification System and set it to score one (1) whenever it is not an exact match (only for international orders). This way you will be accepting international orders only if they have not triggered any checks regardless of the score. Their billing and shipping addresses are the same and it is also the same address that the bank keeps in its records.

Case 4: Chargeback. The customer can be promoted to the suspected fraud database for a period of time or until an acceptable number of new orders are placed without issues. If the customer is already in our suspected fraud database, you should promote it to the negative database.

From some of the results of the different case scenarios, the fraud risk factor also determines the priority of working that order during manual review. Based on our scoring model, an order with a fraud risk factor of two (2) is less likely fraudulent when compared with orders whose fraud risk factor is three (3) or more. Additional elements to consider when assigning a priority on these orders is the date when the order was placed and whether it is an expedited shipment or not.

Also very important is to notice how you can easily tune the model by either changing the score of a simple fraud contributor or changing the threshold. For example, if you want to review all international orders, you should change the score for that particular check from one (1) to two (2).

As you can see, there are different fraud prevention techniques and by putting them in the context of a simple fraud prevention framework it gives you a better picture of what can happen and what to do in that scenario. You should now know how easy it is to adapt to various changes.

eCommerce Best Practices

2.6 Fulfillment / Shipping

2.6.1 Introduction
2.6.2 Importance and Components of Fulfillment
2.6.3 Fulfillment Process in Detail
2.6.4 Challenges and Suggestions
2.6.5 Future

2.6.1 Introduction

A typical eCommerce site contains multiple systems interconnected with each other. Just look at the flow of how a product is bought and delivered to the customer—first, the customer comes to the site and searches for the product he or she desires to buy. The next step is to create an order that contains this product in it. After successfully creating the order, a payment method is selected to pay for the product. After successfully validating that the payment method is a viable one, the order moves into the order management system where a set of business rules are applied to the order. At this point the order needs to move from the eCommerce site to the warehouse/fulfillment. Once the warehouse receives the order, the products are "picked and packed" and shipped to the customer using a predetermined (using business rules) method off shipping.

This chapter will discuss the two parts of fulfillment. In the first part, the fulfillment system from the eCommerce site side will be examined. In the second part, the fulfillment system from the warehouse side will be highlighted. The following will give you insight into how fulfillment/warehouse is attached to the eCommerce site. In order to have successful warehouse integration, the different processes that need to be in place or that need to happen in the background will also be explained.

In this chapter you will also learn about the option of out-sourcing the entire fulfillment process and some of the solutions that are out there for warehouse management. The warehouse management system is not exactly within the scope of this chapter but we will get a brief glimpse into the processes and different solutions that are out there.
Fulfillment from eCommerce Site

eCommerce Best Practices

In an eCommerce site, there are usually two pieces that need to work in tandem in order to make it successful. One part of the site is customer facing and is responsible for collecting the order. The second part of the site processes this order and forwards the order to a warehouse. The latter part is sometimes called the order management system.

In the realm of an eCommerce site, the OMS (order management system) is sometimes referred to as the fulfillment system. There are different processes that are involved in this fulfillment system. The individual processes will be discussed in the following sections and the final subsection will list out the different challenges and suggestions.

Order Validation

One of the first things that a fulfillment system does as soon as it receives the order is to validate it. During validation the different checks that the order goes through are as follows:

- The order contains a billing address, a shipping address, a payment method and items purchased.
- The payment method is valid and has been authorized.
- The tax information is valid.

The purpose of validating an order should be quite evident. This is done so that you bill and receive payment from the proper person, ship the order to the right person and make sure that you have collected the appropriate tax. Validating the order at this point of the process will give you the opportunity to correct any faulty information and reduce any overhead later on.

Payment Re-Authorization

Before an order is sent to the warehouse you need to make sure that you have a valid method of payment. A valid payment method means you have either settled or reserved the funds totaling the order amount from any financial institution of the customer's choice. If for some reason, you are unable to secure funds at the customer's financial institution, you will attempt to secure the funds at a later time for a predetermined amount of time. If you are unable to secure the funds after several attempts then, you should move the order to a separate selection of orders that require special attention.

> **Improve Payment Processes to increase efficiency for you and the customer.**

Fraud Processing

The fulfillment system in an eCommerce site gives you the opportunity to run the order to a set of fraud prevention business rules and determine the level of fraud risk. Based on the calculations you can choose to fulfill the order or move the order to a bucket of fraudulent orders. This bucket of fraudulent orders can be processed manually at a later time. The fraud systems can waver from a simple lookup table to a system with complex business rules. With a fraud prevention system in place, it will definitely help you reduce the risk of fulfilling the order and not receiving funds.

Back-Order Processing

In backorder processing the one thing that you need to remember is that you might have partially fulfilled the order previously. What this means is that any previous authorization on the payment group that you had has already been settled for the goods that you have shipped. This means when releasing the backordered items you must again make sure that you have a valid payment method before sending the item to the warehouse. If the backorder happens every day you must have built-in logic that would check to see if you need to re-authorize the payment method.

Communication Processing

Communication processing means the communication between the warehouse and the fulfillment system as well as communication with any third party vendor involved. In communicating with the warehouse system you will be processing the acknowledgment of the receipt of order to update the order with the shipment information. In communication with third-party vendors you usually deal with any updates to the order that is needed in order to have a comprehensive idea about the state of the order. Some examples of third-party vendor communications are shipment updates from the carriers and the status of a custom-made product from a supplier. It is essential to have a backup for all these communications and also have a business process for what should happen when these communication lines are down. This will result in a more happy and satisfied customer.

> **Improve accuracy of demand forecasting to avoid unexpected purchase periods.**

Analytics Processing

Analytics are quite important for any eCommerce site. The ability to predict the demand for a certain kind of product linked to purchasing trends will help the business in maximizing its profits and at the same time minimizing the overrun costs. It is always good for a business to understand its seasonality—the times during the year that they experience a significant increase in business in terms of volume. For example, it makes sense for an eCommerce site that specifically targets students to be prepared for the back-to-school season. Understanding the customer is a key part of any business enterprise. This same principle plays an important role in a fulfillment system as well. Some of the other analytics that could also be measured is how the eCommerce site is performing at given times in a year as well as under different loads. This helps planners make decisions on upgrading the system.

Inventory Synchronization

Inventory synchronization is important in any retail business. Having correct inventory numbers will enable you to collect valid orders. Having the right inventory numbers will help you reduce cost overruns from backorder processing to inventory holding costs. Usually an eCommerce site maintains its own inventory and the warehouse maintains its own inventory. At a certain point every day these numbers become synchronized. Depending on the business model and the business rules, one of the systems becomes the master system. The master system provides its inventory numbers to all the other systems. Some of the challenges that eCommerce retailers face is the synchronization itself. Ideally it would be nice to synchronize the inventory numbers during non-busy hours. As you can see this poses a challenge as an eCommerce site is open 24 x 7 x 365.

Challenges and Suggestions

Some of the challenges that any eCommerce fulfillment system faces is minimizing the downtime and increasing the throughput of the orders. The former can be achieved by maintaining the systems properly. For increasing the throughput of the orders you have to balance how you are going to implement the different subsystems mentioned above. On average, if an order takes a minute to go through the fraud processing system then the most you can process is 60 orders in an hour. Despite the fact that the fraud system is doing an excellent job, the overall

efficiency of the site has gone down. So it is always necessary to strike a right balance between the different processes that have been implemented versus an efficient fulfillment system.

2.6.2 Importance and Components of Fulfillment

This section deals with the importance of fulfillment systems. Early in the dot-com era, a lot of companies set-up their online presence for failure because they did not think through all the processes involved in fulfillment. Most companies that were successful were the ones that already had a catalog business. The reason for their success was because of a well-established fulfillment system. You have to understand that when dealing with fulfillment systems, you also deal with the logistics of shipping, maintaining inventory, procuring, etc. The companies that failed did not realize this importance.

It is time to look at the different parts of a warehouse/fulfillment system. You now have a well maintained eCommerce site. This site is accepting orders from customers. Now that you have the orders in the system it is important to ship it to them (fulfill). Only after fulfilling an order can we collect the funds from our customers. This presents the first problem that needs to be solved. The solution of this problem is the first piece of the fulfillment system, i.e. the communication channel between the eCommerce site and the warehouse/fulfillment center. Once the channel has been established the second thing that you need to think about is the protocol of communication between the eCommerce site and the warehouse. The third piece of the puzzle is the warehouse itself. For managing a successful and profitable warehouse is to manage not only the space but also the flow of products once they have been picked to where they have been packed and loaded to be shipped. To do all of these aforementioned tasks there are systems designed by companies like Manhattan Associates, Sterling Commerce, RedPrairie, etc. Completing the fulfillment system is the final piece of the puzzle in regards to the backend accounting processes.

Communication

For communicating between the eCommerce site and warehouse there are many solutions. You can set up the communications using a simple socket communication, different JMS solutions like MQSeries and Web services. In simple socket communication, the eCommerce site can open a socket on the warehouse management system and transfer the order for anyone off the text, XML formats. In a JMS solution like

eCommerce Best Practices

MQSeries the order will be serialized and sent across to a remote queue that is connected through a communication channel. In a Web services based communication, the order is sent to a remote site, probably residing in the warehouse environment.

Communication Protocol

Communication protocol is the agreement between the eCommerce site as the warehouse and to how each will respond when a message is received. There are a few simple sites that are basically post orders to a fulfillment site and will not expect to receive any acknowledgment of the message that was just sent to them. There are also sites that require having every communication logged and acknowledged by the warehouse. As seen in the two cases here, the communication protocol could range from very simple to detailed and complex. It is imperative that a protocol is chosen. Once the protocol is chosen it is up to the implementation team to make the protocol work.

> **Better communication with all partners will help iron out disputes and inconsistencies.**

Warehouse Management Systems

The warehouse management system (WMS) is in and of itself a huge topic. A warehouse management system deals with the following systems:

- Logistics of procurement, distribution and shipping: this deals with when to buy the inventory, where and how to ship the inventory.
- Inventory management: this deals with maintaining information about the units of items present in the warehouse or at an off site.
- Slot management: this deals with managing where an SKU needs to be placed, in order to make the picking and packing process more efficient.
- Receive and Respond to the orders that are being sent from the eCommerce site.

Some of the big players in this area of the market are Manhattan Associates with the PkMS system and Accretive with their DmMS system.

eCommerce Best Practices

Back-end Processes

Like mentioned before, Backend Processes are the final piece of the warehouse processing System. This is a generic term that is used for any and every process that is involved in the reconciliation of information between the eCommerce site and the warehouse/fulfillment system. These could involve nightly updates from corporate systems to warehouse and eCommerce sites. These could also involve inventory reconciliation between the different sites, etc.

2.6.3   Fulfillment Process in Detail

There are different steps in the order fulfillment process, which involves not only order placing but also returns and exchanges. The next section deals with the different challenges that are encountered in fulfillment and some possible solutions.

Order Place

The following are the steps in order placing. Keep in mind that this is about eCommerce sites and not traditional order fulfillment processes.

- Order is placed on eCommerce site.
- Payment for the order is cleared.
- Warehouse receives the order.
- Packing slips get printed.
- Pickers pick the product.
- Arrange shipments.
- Packages are shipped to customers.

In the above steps the picker can only pick if the product is available in the warehouse. This brings in the process of inventory management as a precursor for successfully completing the steps off the order place scenario. Arranging the shipments and shipping the products also brings the logistics of shipping and business processes, such as agreements with the shippers or insurance against lost products, into warehouse management. Some of the organizations have a separate department that deals with loss prevention.

Returns Processing

The fulfillment process also has to take care of returns. This process is initiated by the customer when the customer returns the product they received. The following are the steps involved in processing returns:

- Returned package is received in the warehouse.
- The product is returned to its location in the warehouse.
- Initiate payment return process.
- Log the reason for return.

It is imperative that we log the reasons for the return of any product. You will see in the later sections as to why this plays an important role. The reason for returns could be because of "Damaged Package", "Item Defective", "Poor Service", "Poor Quality", "Not as Expected", "Undeliverable", "Wrong Item Sent".

Exchange Processing

The exchange process is basically a combination of returns process and order place. The item that was returned goes through the regular returns process and the exchanged item is placed as a new order. It is a business decision as to how the payment will be handled. The most commonly seen method is to use the refunded amount and charge the customer for the difference.

2.6.4 Challenges and Suggestions

In this section we will use each subsection for one problem and suggests ways to overcome it. At the end of the section, these challenges will be highlighted because overcoming them can translate to a successful implementation of the fulfillment system.

Misunderstanding of Business Processes

A fulfillment operation involves suppliers, shippers, finance institutions, warehouse, etc. Each of these entities has their own language and terminology. Often there is a mismatch between how a supplier handles the product and conveys the information to the warehouse. There could also be similar miscommunication between the warehouse and the

eCommerce Best Practices

eCommerce site. The said business processes could include inventory management, product lifecycle, product replenishment, shipment, etc.

The only way to overcome this is to have a *well-documented business process* that is shared between all the parties involved. The warehouse operation can only become successful if all these business processes are planned ahead of time and agreed upon by all the parties involved. It would be very helpful if the terminology and lifecycles is consistent. Having a good solution for this problem often will overcome some of the other problems like the shipment of wrong products, quality problems, etc. This usually is not a problem for most of the companies that already have a catalog business.

Inventory Problems

Some eCommerce sites have lagged between accepting orders and fulfilling them. This often leads to an inventory problem. The problem could arise because of the forms of payment that are being accepted on the system or because the business is more involved in custom-made products. In any case, some of the inventory problems encountered are cost of storing excess inventory, customer dissatisfaction due to lack of inventory and products being backordered.

The first step involves *product planning*, for any Internet retail site the merchants who procure products go through an extensive analysis of consumer demands and although this is not an exact science, a good merchant can always predict to a certain degree of accuracy what the demand is going to be. This often is the biggest help in reducing inventory problems.

If an eCommerce site accepts checks as a form of payment, than this automatically adds a lag in payment processing which could lead to inventory holding costs. Favoring *electronic payments* will often lead to expedited payment processing and lowering of inventory holding costs.

Sharing the order data between all partners involved in the warehouse operation will also lead to efficient inventory management methods.

> **Manage carrier costs because this can affect the price of merchandise.**

Part II: Selling

Carrier/Shipping Costs and Issues

Carrier or shipping costs are another major component in warehouse operations. Although most often these costs are passed down to the customer, there are also costs involved in getting the merchandise from the supplier to the warehouse. Sometimes, it is not just the cost but also the speed and distance at which the items have to reach the warehouse or the customer.

Most successful operations are those that partner with major carriers who have a lot of expedience and experience in moving merchandise from point A to point B in a given amount of time.

Loss Prevention

When you have a warehouse operation where the people involved are part-time workers or you have a high attrition rate, there is always a risk of theft or damage to the products. Sometimes, this could lead to high insurance rates and loss in revenue.

Some of the effective companies have a separate loss prevention department that tracks losses due to theft or damage and tries to come up with remedies for some of the common problems. Having a department like this will definitely improve the efficiency of the fulfillment process.

> **Consider outsourcing fulfillment and learn the processes involved.**

One other solution that has not been talked about is outsourcing the entire warehouse/fulfillment process. Some of the players in this area of the market are EDS solutions, Accretive, etc.

2.6.5 Future

This section will briefly talk about some of the future developments in warehouse management. One area that is gaining momentum is implementation of RFID technology in warehouse. All items packed are automatically tracked within the warehouse using the RFID tag. This will effectively address loss prevention as well as inventory management. Some of the companies that have finished the implementation of RFIDs are Metro Group, and International Paper to name a few.

eCommerce Best Practices

2.7 Inventory Management

2.7.1 Introduction
2.7.2 Balancing Overstock and Under Stock Items
2.7.3 Suggested Methods

2.7.1 Introduction

Managing your inventory is an essential function of an eCommerce site, especially if you run a multi-channel operation. Keeping track of what you sell versus what is available to sell is basis of any business. Customers will assume the products they want to purchase are in stock unless told otherwise. Synchronization between offline and online inventory management systems requires accurate back-end tools that are able to provide up to the minute data feeds.

Even if customers are disappointed when shown that a product is out of stock, at least they were notified early in the process. Customers want to be informed and aware of the situation before committing to making a purchase. If a customer purchased an item and later gets notified that the product is out of stock, then he or she would have to go out of their way to remedy the situation. Avoid these mistakes by fine-tuning your inventory management.

Having a solid inventory management system in place will reduce the risk of complications. A good system will be able to forecast and track overstocked items, as well as detect when items are out of stock.

2.7.2 Balancing Overstock and Understock Items

Inventory management is tricky because retailers often see that when product demand increases, availability also increases. The immediate reaction from sales is to over compensate and increase purchase orders, but what happens if demand suddenly decreases, and you are left with extensive overstock? Backordered items give customers time to seek another source and you want to avoid that if at

> A balance between overstock inventory and outages will let you know what products to push and what needs to be re-stocked.

eCommerce Best Practices

all possible. Some of the methods to have a balance between overstocks and outages are:

Identify the Percent of Inventory

Retailers should focus the bulk of their efforts on items that yield the greatest results, meaning always identify the top twenty percent of your inventory which produces the most sales, and prevent your sales and warehouse teams from letting a backordered request to pop up.

Service with a Risk

Retailers should question and identify the risks associated with each purchase. What is the cost if item 'X' is backordered? How much would the overstock end up costing?

Tie-up with Warehouse

Retailers need to negotiate the cost of goods leading to packaging and delivery and extras. Inventory management should be linked-up with warehouse management to maximize efficiency.

Plan and Management

Whenever there is an increase in sales, retailers should leverage efforts towards planning and managing inventory carefully so that incase there is a sudden dip in sales the next week, the overstock dilemma does not come to fruition. Retailers should minimize the number of warehouses maintaining inventory that is not moving during a sudden decrease in demand of a particular product.

Liquidation Process

When an item is removed from the catalog, the liquidation process must start with a pre-laid out plan to generate the maximum amount of returns.

2.7.3 Suggested Methods

With the advent of new tools in the market today, sales and marketers are shown a greater degree of flexibility that can be attained with the implementation of behavior targeting—one of the more effective aspects of

eCommerce Best Practices

inventory management. Sales teams have to know what inventory they have available by site and also by audience type.

Behavior Targeting

When sales people sell a behavior targeting campaign, sometimes they do not know and cannot anticipate the impact on their inventory scale. Thus, preparation is needed to understand the scope of the campaign. Tools are available in the market today that can help retailers build a management process that allows sales people to tie both ends together and provide an accurate feed to customers' shopping patterns.

> **Behavior targeting helps retailers and sales teams anticipate the impact of preplanned inventory.**

Know your Inventory Tool

The current applications available today are more sophisticated than ever before. They incorporate sales plans and inventory replenishment modules which produce detailed information of forecast inventory management modules. You should make sure that your sales team is trained in interpreting this type of information and module, and be able to act on them to get the desired ROIs. Tools are divided and dependent on whether you carry small or big inventories. Small retailers should know the limitations of these tools and base their inventory plans for effective management.

Understand Customers

Retailers and marketers should understand their customers' needs and expectations. The collected data should factor in shopping patterns from the past and current trends to accurately forecast customer demand and plan inventory levels to meet this demand. Seasonal progress should be observed in order to help retailers maintain an inventory level that minimizes markdown exposure and updates plans as the season progresses.

Controlling Inventory

Good inventory control is critical to any retailer as it ensures that an adequate level of stock is available on hand for the correct amount of sales being made. Sometimes retailers have too much inventory and this

eCommerce Best Practices

slows down the cash flow, forcing you to strategize plans in overcoming it, usually by introducing sale scenarios that are not dependent on season. Shopping data collected from the store level can be leveraged as real time data. This is vital information that needs to be shared across the retail supply chain network. It will ensure that the chain network consisting of suppliers, factories and distribution centers use this information to meet anticipated product demand.

Bulk Inventory Distribution

Some retailers are large in size, typically storing and receiving massive amounts of inventory at their distribution centers. These bulk inventories are sometimes broken down and shipped to different locations to replenish the sold merchandise. This is more costly and time consuming—why? Because manual jobs require a lot of labor and man hours. You would have to then receive these bulk inventories, open and repack them into smaller units, and then send them to stores. You should avoid this time consuming and costly process by turning to methods that bypass these steps so that they are received at the stores themselves based on accurate replenished data.

Counting Accuracy

Some retailers shut down their warehouse operations to take a complete inventory management check. To avoid this, retailers are implementing a more effective method of cycle counting where they take stock of goods on an ongoing basis—some based on frequency, for example, like the best selling items grouped under category A are counted more frequently than the average selling products under category B, while the marginally sold and occasionally counted products are listed under category C. By tallying under such categories, retailers have a better control of handling what is in stock, and they can cut on their backorders, boost first time fill rates and improve customer satisfaction.

> **It is important to accurately count and track products, which can be accomplished by categorizing and organizing based on frequency.**

Promotions Require Synchronization

The creation and development of marketing plans require close communication with the creations of merchandising plans in order to produce successful strategic inventory management. You should plan out a calendar and set projections for all promotions in channels like catalog, media, email, and retail stores. Merchants and inventory control groups should plan product purchasing, and availability to support these events. You should ensure that during this planning there are three aspects in conjunction with one another—marketers compiling and updating the marketing timeline, retailers planning expected orders by week for all promotions, and the inventory team forecasting the demand for these promotions.

eCommerce Best Practices

2.8 Merchandising

    2.8.1   Introduction
    2.8.2   Effective Techniques Used
    2.8.3   Recommended Methods

2.8.1 Introduction

Most retailers can merchandise their products effectively in their showrooms but find it difficult to replicate the same concept online. Sometimes online version of the showroom is successfully reproduced, but garners little to no returns when compared to their offline methods.

Strong Merchandising can make browsers in to buyers. A large percentage of people shopping online do not have an idea of what they want to buy until they see it. Retailers try to target these shoppers by associating what they sell to a certain lifestyle or functional need. There are ways to design your site that is organized and branded so that it sells.

Expert merchandising strategies are needed to keep shoppers engaged. This increases conversion rates and order sizes. Effective merchandising is essentially a confluence of data driven analytics, seasonality strategy, promotional practices, price adjustments, design, and product display and information. It is an all encompassing word that generally means positioning your merchandise in a way that makes people buy it.

2.8.2 Effective Techniques Used

Merchandising is not just about selling your merchandise; it is about how you go about selling it. It is about providing the customer extensive product information, promotions, graphics and search capabilities on your eCommerce site. One way to move from product selling to product merchandising on a web site is to convey lifestyle images that incorporate several of your products.

Some retailers use lifestyle photographs in each section of their site to create a sense of product enjoyment. This can work well with marketers and their different strategies to a certain extent, but beware of savvy consumers that can

see through the images and realize that it is still a product-oriented eCommerce site.

The most successful technique would be if your site takes a step further from the earlier method and create lifestyle sections within your site that has products from different categories grouped under various headings. These sections would have collections that make complete outfits or product packages for men, women and children. Customers might not end up buying this complete collection but seeing how several items work together would make them purchase more than one item.

> **Images that show products grouped as a collection gives your merchandise a frame of reference.**

Retailers can and should enhance product merchandising by using beyond the standard use of photographs of products. Some being:

- Photographs of people using the products that encompass various lifestyle images.
- More product information like detailed product descriptions, how a product or product package can be used.
- Different angles of certain products appeal to the consumer, which should be shown effectively. Also showing it being used in different angles enhances the sale prospect.
- Sometimes several products grouped together as a collection in a photograph or on a model also can work to sell more than one item from it because it gives customers an idea of how it can be used, styled or maintained.
- Zoom-in and out features allow customers to get a closer look at the product details and quality.

2.8.3 Recommended Methods

Consumers want to have control over their shopping process and not experience any pressure from online salespeople. They want room to shop and decide at their own pace how they would like to add, delete or modify their basket items.

For retailers to be effective in merchandising, they need to understand what makes things sell:

eCommerce Best Practices

Searchandising

Having a prominent search mechanism on your site is extremely important as it gives customers the ability to easily narrow down their product selection. Of course this means that there needs to be organized content management and categorically identified products, whether it is based on brand, price, color, function, or product line, etc. A good search mechanism is one that hones in on what the customer is looking for with only a few clicks of the mouse.

Do Not Force Sell

Some retailers try forcing un-popular products on consumers due to overstock or a lull in sales for that particular product. However, this can drive away new users, so you should be careful on how you approach marketing these products to consumers.

Show Best Offer Upfront

Marketers should know that whenever they hide a good offer behind a bad offer, they are limiting their sales potential. Do not give customers an opportunity to think about or second guess promotions. Manage all your active promotions and try not to offer so many as to overload the customer with every single one. Promotions should never be confusing or convoluted, nor should it take precious time away from shopping. Catch consumers' attention with the products they are interested in: this can be attained from user analytics or tracking the user behavior.

Dare to be Different

Promote personalization with each customer in a different way. If a customer is shopping on your site for the first time and buys a bunch of shoes, then you should strategize ways to sell other accessories that could go well with the products, like socks, watches, handbags, etc.

Dynamic Landing Pages

Retailers can take advantage of building a special merchandized results page that is used to promote a brand when people are searching for the same. Dynamic landing pages also help

> Dynamic landing pages can sell a specific brand and be used to target email newsletter recipients who are driven to this page than your homepage.

in email strategies where you can direct users to the brand page rather than the homepage.

Virtual Shopping

Since the web gives more flexibility in the method to present a product to the consumer, retailers should find creative ways to encompass the same. Products that can be shown in detail should be shown so that the consumer gets a real time feel.

Merchandize by Refinement

Retailers mostly apply attributes like brand, size, or price when they group products logically. This assists in increased conversion rates and sales when presented correctly. Merchandising should involve promoting products by refining attributes either globally or locally within their site.

2.9 User Ratings / Reviews

2.9.1 Introduction
2.9.2 What Type to Include in Reviews
2.9.3 Balance of Positive and Negative Reviews

2.9.1 Introduction

Providing user reviews and ratings were initially viewed as as potential threats to sales growth because they exposed your brand to public scrutiny. You cannot stop people from talking, but even more importantly you cannot stop them from listening. Word of mouth and word of mouse can have a viral effect if a review was scathing enough, which is a primary deterrent for retailers considering incorporating this functionality. However, if people want to talk about products they will find a way to do it, whether the merchant facillitates the process or not.

Customers often prefer to shop on sites that offer user ratings and reviews because they want feedback and reassurance that they are making a right decision in buying the product. The ratings also give you information about what products are popular and what products are struggling to find an audience.

> **The growing sense of community in the online world contributes to their sense of duty to inform one another about products they love or hate.**

The online world, not only in eCommerce, is experiencing this shift in the way people interact with eachother. There is a greater sense of community even if members of that community are in multiple countries around the world. Because of this idea of community with one another, shoppers are adopting a self imposed duty to inform other shoppers of their experience with the product. A company's marketing and advertising make shoppers interested in a product, but it is other users that may determine whether he or she buys the product or not.

Shoppers are more likely to trust customer ratings and reviews rather than what the merchant tells them. The reviews also help convert first-time buyers—where a shopper who has not previously bought anything from the site examines the reviews and decides to make a purchase based on the information the reviews gave.

eCommerce Best Practices

Though there is no baseline to measure and quantify customer reviews, shoppers generally tend to be positive and merchants need not fear about unsolicited feedback. Customers embrace, accept and appreciate reviews and feedback on websites.

2.9.2   Products in Review

The types of products and services customer ratings or reviews would benefit the most from is a question you should ask before implementation. Technically anything and everything offered online is reviewable, but there are some product types that can experience more sales growth than others with product or user reviews. Generally, the more important the purchase is, the more emotional the purchase, and if the purchase requires a high investment, the more influential word-of-mouth is going to be.

> **The greater the level of importance, emotional attachment, or investment to a product the more influential reviews.**

Commodities such as sponges or window cleaner may not be the most optimal subject matters for customer-written reviews, but items like refrigerators, televisions, music or hotels are ideal for reviews in terms of the help they would provide when making a purchase. Users are more willing to take advice from people like them rather than from the company that is trying to sell them the product. With customer or product reviews the viewer generally makes the assumption that there is no agenda or ulterior motives to the positive or negative reviews—it is honest feedback that people may or may not choose to listen to.

Understand that user reviews and ratings increase user experience on your site and adds value to your brand. The online market gives people the capability of doing their own research, especially for large investment products like a washer and dryer. If someone is comparing washer and dryers, then product reviews and user ratings on your site can be assets because they would not have to exit your site during the decision making process.

You need to weigh the benefits of providing reviews versus the effort it takes to implement and maintain them. Keep in mind that low-emotion product categories may not be worth your time.

eCommerce Best Practices

2.9.3 Maintenance

Deciding to allow an open forum for potential ridicule and scrutiny is no easy decision, especially because you have no control over what people say or rate, and if you delete all the negative reviews than that leaves your company looking even more suspicious and would defeat the purpose of creating them in the first place.

> **Make sure the reviewer is actually posting a review of the product itself and not any other unrelated subject.**

However, that does not mean you cannot implement rules and controls over customer reviews. Some obvious controls involve users writing profanity or disparaging remarks, mentioning of direct competitors, and most importantly the relevancy of the topic being reviewed. Do not forget that it is supposed to be a review of a specific product—not the price, fulfillment process, etc. Any review on anything other than the actual product itself may be filtered out.

It is important to make the most of the negative reviews. Negative reviews can help you spot and address problems which you may not have seen on your own. There might be enough negative reviews to warrant a change in product design, functionality, inventory levels, etc.

2.9.4 Balance of Positive and Negative Reviews

A mentioned before, removing every negative review will do more harm than keeping it on the site as you call into question the integrity of not just the product, but the company as a whole. If you are a large company the last thing you want to be accused of is tampering customer reviews. You have to be able to take the bad with the good. Customers talking to

> **Do not remove every negative review or post fraudulent positive reviews. You risk damaging your company's reputation.**

each other give you a tremendous amount of insight about who is shopping from your site and what it is they are looking for, so there is an opportunity to learn from negative reviews.

Fraudulent positive reviews are just as detrimental to your integrity as removing negative reviews. These efforts tend to be transparent as customers can usually see through these reviews. Some companies allow customers to flag reviews they

find more negative than it should be. This ensures that they take a look at the particular review and make amends as required. You also should make sure that you do not edit the reviews for spelling or grammar. The context of who that person is will be very important for whether they will believe the review or not. Spelling errors on reviews should be left as is so that it adds credibility and humanity to the person that wrote the review. You do not want a shopper to read a review and think that one of your lawyers drafted it up.

Implementing online customer ratings and reviews is about getting over the fear of losing control. Before, the way customers came to learn about a brand was through the messages that the company decided to feed them. Making them give up that control and letting customers come in an open dialogue about the product is revolutionary. Retailers need to come with terms with customers forming collective online communities that depend on one another for making informed purchasing decisions.

eCommerce Best Practices

2.10.1 Gift Services

2.10.1 Introduction
2.10.2 Some Common Gift Services
2.10.3 Customer Relationship with Gift Services

2.10.2 Introduction

Gift givers are one of the most prominent types of online shoppers. That being said, so few Internet retailers are directing their products or services to people who are not shopping for themselves. So how do you personalize and develop CRM initiatives for gifts and gift givers? There are also different functions you should consider for gift services like the option of gift wrapping with shipments or sending personalized notes to recipients. Equal thought needs to be given to whether the gift products can be delivered to a different address apart from the existing consumer—some payment processes require the billing address to match the shipping address. The Holiday season is definitely the most important and lucrative time periods for retailers looking to leverage their gift services to loyal customers and new users.

> **Create a space on your site dedicated to gift seekers, and be sure to leverage your top rated products to drive more gift sales.**

Designate an area on your site for people searching for gifts. You can organize the gift recommendations by user ratings, top-sellers, price, critics' picks, etc. In an eCommerce shopping environment that is greatly influenced by user ratings or popularity, it is surprising to find that not many retailers are directing that knowledge towards gift givers.

You may need to strategize various methods that build relationships using gift center ideas in order to drive new users to the site. Some retailers aim to target shoppers during special occasions like Valentine's Day, or Mother's Day. The returns are great if the marketing strategies are executed well. Gift services could be an added benefit that may increase sales and are good ways to encourage repeat purchasers as well as attain new loyal customers through the gift recipient.

> **Brand your gift services to ensure customer quality and detail—something you should not compromise.**

eCommerce Best Practices

2.10.3 Some Common Gift Services

Because of the long-tail nature of the Internet, people are able to fill voids in the market that were previously un-tapped, and due to the multitude of products available, the opportunity for gift services seems only natural because no one can possibly buy all those products for themselves.

The idea of gift giving is nothing new, but unlike the traditional process of exchanging gifts, people using online gift services are leaving a lot of the responsibility to the retailers in terms of packaging, gift wrapping, and safe delivery. Gifts are associated with a number of types of services, some of which are mentioned below:

Gifts with Product Purchase

One of the main issues with online gift giving is delivery. If someone buys a product online they want reassurance that it will be shipped to the right address. Many shoppers want to gift wrap their purchase and may need to send it to an alternate address or even multiple addresses that might not be in their list. Back-end systems should accommodate sending gifts to the correct addresses in order to ensure customer satisfaction, especially if you are offering a gift wrapping service. Retailers need to tackle such unique delivering patterns so that intended person receives the gift.

Wedding Gift Services

Guests invited to weddings are one of the most common and much targeted gift service users. Prior to the wedding date they frequent retail sites, some of which have a separate section dedicated to wedding gift services. Customers who are not able to attend the wedding find the separate service a useful and convenient method to send items to the husband and wives to be's. This is commonly known as gift registries, which hold a listing of all the requested gifts invited guests can purchase for the newlyweds.

> **Gather data which would segment the type of customers looking for gift services on your site.**

Retailers should notify shoppers viewing the registry with key information like what items have already been purchased, cost of each item, how well in advance they need to purchase and send items, limitations, if any, with respect to location and shipping costs, etc. Some gift services let wedding guests buy one part of a gift set, for example, one pot or pan in a whole set of pots and pans. This allows guests to budget their gifts, yet still give the bride and groom what they want.

Gifts for Season or Occasions

Some retailers only implement gift services during certain times of the year like the holiday season, during the back-to-school season, or for special occasions like Valentine's Day or the Fourth of July. Retailers try to make most of their sales during these time periods and their marketing strategies should be executed in a timely manner (typically one or two months prior). Sometimes retailers choose to not offer gift services during the regular shopping time period to cut down on costs.

However, an equal number of retailers offer gift services during season and non-seasonal times. As a retailer you need to decide whether your products have the flexibility and need to offer gift services year-round.

Charitable Donations

Charitable organizations that accept donations online can be considered under the gift services umbrella. A donation can be made on your behalf or on someone else's and instead of receiving a commercial product to "unwrap", typically a confirmation or Thank You letter is sent via email, or even a certificate of donation can be sent via snail mail. Some organizations even let them track where or whom your dollars are being allocated to. This, for one, shows that their money had a purpose; secondly, it gives the donator reassurance that it is a legitimate charity; and thirdly, encourages future donations because it gives a "face to the name" so to speak.

Another way charities are acquiring donations is to act like traditional eCommerce sites, in that they offer products for people to buy, in which the money or a percentage of that money goes to the charities. Anything from celebrity auctions (when people buy products worn, owned or designed by their favorite celebrities) to T-shirts that promote the cause can be bought online and used towards charities. In addition, people can buy products that serve a charitable function, for example mosquito nets for children in dense Malaria stricken areas.

2.10.4 Customer Relationships with Gift Services

Loyal shoppers and new customers should be placed on a customer relationship platform that drives genuine satisfaction and loyalty for effective gift services. This injects brand and product exposure with equal measure. Some of the standard suggestions to consider are:

> **Know the customers that are generating sales through gift service usage. Target them with more incentives that drive sales.**

- Do not skimp on customer service. There is a good chance many people buying gifts will contact you about their order, whether it has to do with gift wrapping or failure of delivery. It is important to have exceptional customer service during the holiday season because it usually involves two different parties (the gift giver and gift receiver). Making a bad impression here is too risky.

- Promise what you can deliver. With the increasing number of retailers offering different types of gift services, retailers should view customers as highly influential, individual brand ambassadors, in which their word-of-mouth increases your market share. For this you should ensure that the services your site promises to deliver are of quality. This leaves customers satisfied and increases the likelihood they will return or send referrals to friends and family

eCommerce Best Practices

2.11.1 B2B

2.11.1	Introduction	
2.11.2	Attributes	
2.11.3	Usage	
2.11.4	Functionalities	
2.11.5	Technology, Standards, Considerations	
2.11.5	Functions and Tools support	
2.11.6	B2B	
2.11.7	Collaborative B2B	
2.11.8	Benefits	
2.11.9	Recommendation	

2.11.2 Introduction

Business-to-Business, or B2B as it is commonly known, is the term used to clearly distinguish a specific class or type of eCommerce. B2B refers to the activities and transactions between business enterprises while the more familiar Business-to-Consumer, or B2C, refers to transactions occurring between a business enterprise and consumer (the general public).

> **B2B has great potential to cut down costs, increase efficiency, and realize faster time to market.**

B2B eCommerce provides back end support for sustaining growth and adoption of B2C eCommerce. Some of the conceptual and technological foundations for B2B were put in place in the form of earlier technologies like Electronic Data Interchange (EDI). However, the real growth and adoption of B2B may be attributed to the proliferation and maturity of the Internet and its associated standards and technologies.

> **Processes of B2B eCommerce make transactions more convenient as they allow for order automation based on inventory or seasonality.**

B2B eCommerce has streamlined the processes between business enterprises. Because B2B transactions occur on a more regular basis than B2C purchases, many of the orders can be automated to re-stock inventories or fit seasonality.

This class of eCommerce may take varied forms and formats, even though some standards have emerged and are constantly emerging. This chapter will make

an attempt to find common threads among the variations and suggest some best practices for the adoption and usage of the technology.

2.11.3   B2B Attributes

Some basic characteristics of B2B eCommerce can be summarized as follows:

It occurs between business enterprises.
It generally entails a large volume of business transactions.
Both performance and usability are important for B2B; however performance generally takes higher precedence.
It has objectives of reducing overall cost, offering value-added services, making efficient business operations, building brand image, etc., which collectively has a long-term goal of higher profitability.
It generally entails order approval processes in the buying entity, acquisition of orders in the selling entity, some form of contract or intermediary exchange party, and order fulfillment processes.

2.11.4   Usage

Varied usage of B2B eCommerce has been observed. The variations cater to different economic sectors, though they themselves can be in different stages of evolution and can cater to any particular sector in different manners. Some important ones are listed below.

In some instances, B2B eCommerce activity may take a vertical market approach where it focuses on integrating and aggregating outputs from the same industrial sector.
In some other instances, B2B eCommerce activity may take a horizontal market approach where it entails integrating outputs from different industries and its sectors, thereby building collaboration across different fields and enabling you to diversify.

Still, it may take a mix of both horizontal and vertical market approaches.
Some of the B2B eCommerce partners may employ a pay per use scheme for services they offer to others while other partners may use subscription/membership/license fees at a flat rate. Determining which one is appropriate depends on the volume of transactions, frequency and time period for which the services are required or are being offered.

Various eCommerce platforms and tools have come up in the market and some of them provide reasonably complete suites of tools for B2B.
B2B eCommerce may go beyond just buying and selling and may address aspects like supply chain management.

2.11.5   Functionalities

There are a number of functionalities that you might want to include in a B2B eCommerce application which are typically not implemented or used in a typical B2C site. These are:
- Contracts,
- Custom Catalogs,
- Custom Price Lists,
- Scheduled Orders,
- Roles,
- Approval Process,
- Multiple Shopping Carts,
- Invoicing Support

Custom Catalog and Custom Price Lists allow you to establish a custom relation with your business contract, as specified in the Contract. Since each contract has separate conditions and agreements, custom catalogs allow you to only show those items that the client can buy. Custom price list also provide similar functionality as in showing only the client specific pricing.

Scheduled Orders. You might want to implement scheduled orders for your B2B site if the client has a need for recurring orders. This gives the client a dependable and reliable way to reorder frequently used supplies.

Organizational Support: In B2B sites, the user is typically associated with the company they belong to. The need for organizations as a concept in your B2B site becomes important as you might have multiple people from the same site acting in different capacities or Roles. A person with an admin role might go in and edit other people within the same organization and assign them different roles to serve within the context of the B2B site.

Approval Process: Since large orders might be placed on behalf of the client/organization, these might require an approval before they can be sent for fulfillment. This approval can be at the client side (via the B2B site itself) or from your company's perspective

Multiple Shopping Carts: If an order has to go through an approval process, then there is potential for a wait time before the next order can be placed. This can cause an unnecessary delay in the organization, especially if the next order might not require approval. The client should be given the ability to have multiple shopping carts to prevent such processing delays.

Invoicing Support: In addition to being able to pay with a credit card, the B2B customer should also be given the option to have the order invoiced by your company. This might be because the B2B customer may restrict the usage of corporate cards or if their existing processes

2.11.6 Technology, Standards, Considerations

There are certain important considerations about B2B eCommerce initiatives such as a comparatively larger infrastructure, involvement of many participants, rules that span a single company, large volume of transactions, more orchestrated business processes, need of certainty in order fulfillment, etc.

There is an increased scope of data security and standard requirements. Some standards like EBXML (Electronic Business using eXtensible Markup Language) have come up recently with the objective of providing an XML-based open platform for realizing interoperability.

There are various layers of common specifications required for Retrieved from "http://en.wikipedia.org/wiki/Business-to-business"
Business processes, messaging, collaboration and protocol, etc. Some of the emerging standards are ebMS (ebXML-messaging specification), ebBP (Business Process & Collaboration), and CCPA (Collaboration Protocol Profile and Agreement).

2.11.7 Functions and Tools for Support

Various functions of B2B eCommerce can be summarized as followed:

- Purchase and management of procurement.
- Management of supply and supply channel.
- Management of inventory.
- Management of sales and payment.
- Management of services and support.

eCommerce Best Practices

Software tools are valuable and available for various functionalities of B2B eCommerce, including:

- Purchase orders.
- Requests for quotations (RFQs) generation and response.
- Catalog creation and management.
- Processing of invoices.
- Online negotiation.
- Dynamic pricing.
- Transactions support.
- Procurement and Inventory management.
- Bill of materials generation and processing.
- Supply and supply channel management.
- Sales and payment management.
- Accessory services and support management.
- Integration with back end systems like ERP, CRP, Inventory Management System, planning and forecasting system, etc.

2.11.8   B2B Exchange

Fundamentally, B2B exchange is a web site acting as a kind of virtual market where various companies can buy and sell their products and/or services. Some exchanges also offer additional or value-added services.

The exchanges can be publicly owned with a high degree of participation from various sides and industries in strategic decision making or can be privately owned by a single industry. Which one to choose is determined by the type of products or services required or offered, security considerations, relationship with exchange in question, etc.

If you are a buyer it is a good idea to use public exchanges so you can see different competitive prices and if you are seller it is a good idea to be on a public exchange in order to increase customers. For previously established transactions and if you have a relationship with a private exchange that has a high degree of credibility and security/privacy assurance, the private exchange could be the better choice in this case.

Exchanges also differ in terms of how they are used or are operated. Some exchanges provide platforms and mechanisms for multiple buyers to build an economy of scale in their purchases, while others provide an opportunity for sellers from different industries to create an aggregated sales system in order to cross-sell or up-sell their products.

2.11.9   Collaborative B2B eCommerce

Collaborative B2B eCommerce is going beyond just buying and selling. It entails establishing a more intimate and encompassing relationship. It generally involves allowing access to internal systems and/or processes in real-time and may attempt to integrate supply chain.

> Use B2B not just for buying and selling but consider making it as valuable platform to improve efficiency and overall profitability.

A collaborative eCommerce can foster more efficient business operations, faster time to market, increased responsiveness, and productively managed inventory. However, building it is full of challenges as it requires commitment to make up-to-date reports and share their data with others from involved parties to benefit.

2.11.10   B2B Benefits

B2B may provide several benefits—both tangible and intangible. Some of them are:

- Increased responsiveness to customer demand.
- Reduced price.
- Enhanced competitiveness and faster time to market.
- Reduced paperwork.
- Elimination of geographical barriers.
- Efficiently managed inventory and elimination of unnecessary inventory build-up.
- Enhanced information flow and transparency.
- Reduced processing and interfacing costs.

2.11.11 Recommendations

- B2B is a great platform for enterprises to engage in buying and selling but its scope can be beyond that. It can be used to realize efficient business operations, faster time to market, efficient inventory management, etc.
- EDI can cater to some of the requirements for B2B eCommerce but has limitations in terms of scope and standards. Use open standards to realize extensibility and more potential benefits.
- It is important, however, to determine what degree and type of eCommerce setup would be appropriate for companies based on current and envisaged requirements.
- The progression of eCommerce setup can be considered as starting from creation and placement of an online catalog. The next step may consist of creating a portal for partners that will integrate with your back end system. Realization of the second stage may take longer, but it can provide more benefits. Mapping business processes with other partners and realizing collaborating eCommerce is the most time consuming yet most promising type of B2B eCommerce.

 eCommerce Best Practices

2.12 Soft Goods

2.12.1 Introduction
2.12.2 Differences Between Soft and Hard Goods (Consumer's POV)
2.12.3 Major Differences Between Soft and Hard Goods (Provider's POV)
2.12.4 Verticals that Deal Mostly in Soft Goods
2.12.5 Dealing with Piracy and Unauthorized Use of Soft Goods

2.12.1 Introduction

Soft goods, as the term implies, are products that lack durability or solid shape. In the conventional commerce world soft goods traditionally mean textiles, linens, clothing, etc. However, in the eCommerce world the term refers to products that are totally intangible, delivered primarily by electronic means.

Soft goods include such things as digital music, written content, digital video, software applications, digitized pictorial content, etc. The term also refers to merchandizing items such as digital coupons, e-cards, and community memberships. This merchandise is provisioned for an increasing, perhaps exploding, number of devices from personal computers to mobile devices, to set-top boxes.

Soft goods, while presenting a number of opportunities to retailers, also present a number of challenges unique to their nature and constitution.

2.12.2 Differences between Soft and Hard Goods (Consumer's POV)

The major difference between conducting eCommerce in soft vs. hard goods is in their provisioning. Once purchased, hard goods require physical shipment of the merchandise, selecting a method of shipment, providing a physical address, and possibly paying for the delivery service. Furthermore, with hard goods, the time between when the order is placed and the time at which the merchandise is actually delivered can range between a day to even several weeks, depending upon its stock availability and the method of shipment. Almost always, actual debit of the consumer's account for the cost of the merchandise is deferred until the merchandise is physically shipped, as dictated by law or by best practice.

Part II: Selling

eCommerce Best Practices

Provisioning of eCommerce soft goods, on the other hand, is much more immediate, delayed only by the bandwidth of the consumer's internet connection. Payment in most cases is also immediate as the soft good is either delivered electronically or available for download at the time of purchase. There is also no need to choose a method of shipment—other than perhaps the selection of a download site—and no charge associated with the delivery. In many cases the consumer is required to provide an email address, but almost never a physical one.

The lack of a requirement to provide a physical address when purchasing soft goods has a subtle but significant implication for the consumer. He or she may perceive a sense of privacy that is not present when a physical address must be provided for delivery of the merchandise. This difference, while seemingly insignificant, is perhaps responsible for the astronomical growth of entire industries such as internet pornography, where this sense of privacy may remove an obstacle or hindrance to the transaction.

Another difference between hard and soft eCommerce goods is the way in which the consumer is able to sample the goods. While purchasing goods in a traditional market place—the so-called brick-and-mortar stores—the consumer has the opportunity to sample, touch, and view the merchandise first-hand. In the eCommerce market place, however, hard goods can only be sampled by viewing an image, and to even get the most real virtual experience of the product you would need RIA capabilities like 360 degree views, close-ups, etc.

> **Customers are able to sample soft goods, unlike hard goods. Being able to "test out" files is one major benefit of soft goods, which also leads to higher conversion rates.**

Soft goods, on the other hand, by virtue of their digital constitution can be sampled exactly as they are provisioned, with perhaps some limitation imposed by the provider on their completeness—for example only part of a song or excerpts from a publication—or sometimes some artificial alterations such as watermarks on images to prevent piracy.

eCommerce Best Practices

2.12.3 Differences Between Soft and Hard Goods (Provider's POV)

As with the consumer, provisioning also presents the major distinction between soft and hard goods for the provider. Hard goods eCommerce typically requires much more elaborate integration with back-end systems to acquire inventory status, fulfill the order, track shipment, manager back-orders, etc. Soft goods eCommerce still requires sophisticated content management systems to track, ingest or procure, and categorize merchandise, but the tasks of inventory tracking and order fulfillment are greatly simplified by the reproducible nature of soft goods.

This reproducible characteristic is also one of the major challenges facing the provider and/or producer, as it makes it difficult to prevent piracy and unauthorized use. The difficulty lies directly with the digital nature of soft goods, but technologies and strategies have emerged to help providers deal with the situation. The technologies and strategies will be discussed separately in this chapter.

2.12.4 Verticals that Deal Mostly in Soft Goods

While most enterprises that engage in eCommerce usually make use of e-coupons and other soft goods, some industries deal exclusively in them. At the forefront is the online music industry which appears to be redefining the music industry at large as an increasing number of people—particularly the youth culture—are opting to download individual songs, and sometimes illegally share them, instead of purchasing the album in a hard good format such as a compact disc. The implication of this trend, the popularity of the iPod™ and other MP3 players, and the emergence of controversial sites such as Napster have the traditional record labels and producers whom own much of the content scrambling to restructure and redefine their business model in order to partake in the revolution at hand.

2.12.5 Dealing with Piracy and Unauthorized Use of Soft Goods

Soft goods are easily reproducible and copied by their very nature, presenting a challenge to suppliers as piracy and unauthorized use can represent a significant drain of potential revenue.

Part II: Selling

The ability and ease of which soft goods can be exchanged has even spawned whole new industries that facilitate it by providing archiving, taxonomy and repository services, prompting intellectual proprietors to seek relief from the legal system in order to stem the tide. Napster.com once emerged as the premier MP3 and Audio Download Services website and was faced with a slew of law suits from content owners seeking to thwart their attempt to become a destination where digital music consumers go to exchange their files without having to pay royalties to their intellectual property owners. Under this model, Napster would generate revenue mostly from advertisements. The courts sided with the property owners and issued injunctions against Napster, forcing it to change its business model.

While legislation is one strategy for dealing with unauthorized use of soft goods, it is not well-suited to deal with all the individual violators as well as it can deal with businesses like Napster that attempt to exploit this soft good vulnerability. In addition to that, record labels and entertainment executives are wary of upsetting mass groups of consumers that may boycott their brand in protest of their involvement. This is not the sort of press they want to garner or call attention to. To that end, several strategies have emerged belonging to the domain of Digital Rights Management.

One of the most ubiquitous schemes is the one adopted by Adobe's Acrobat reader which has facilities to prevent the contents of a document from being printed. Another permission management scheme is being used more and more by software companies that embed a requirement for registration in their applications. This registration engages a tracking system which manages the rights associated with the application. A special case of this scheme is pay-per-use where registration (implicit or explicit) is required with each use.

Apple's iTunes™ has arguably been the most successful soft goods retailer in terms of balancing the entertainment industry's issues of piracy and the consumers' increasing digital demand. Instead of charging a registration fee, iTunes™ has set-up a pricing system where most of their songs can be purchased for $0.99 each or the whole album for $10.00. Consumers do not have to worry about facing any legal issues and record labels can finally rejoice in enjoying the kick-backs off the digital revolution. Because of its popularity, as well as influence, the iTunes™ music repository is constantly growing and is now considered a vehicle to jump start up-and-coming artists. iTunes™ is so successful in the music industry that television and motion picture industries are becoming more and more involved as well.

Soft goods comprise a sizable percentage of eCommerce. Some industries deal exclusively in soft goods, while others use e-coupons and e-cards as incentives to increase revenue in their traditional hard good eCommerce businesses.

Commerce involving soft goods affords simplifications in some areas and complications in others in the implementation of eCommerce solutions, but the growth and expectation of the marketplace as to their availability make it mandatory for online merchandisers to make provisions for them.

New standards and technologies are emerging to deal with the management of intellectual property rights, an issue often associated with soft goods, stemming from their digital constitution. These standards and technologies will become increasingly important as the number of digital assets and devices to which they can be provisioned proliferate. Marketers wishing to exploit the rapidly-growing marketplace of soft goods should be well prepared in order take advantage of these opportunities and to deal with the challenges specific to this class of merchandise.

eCommerce Best Practices

2.13 Gift Cards

2.13.1 Introduction
2.13.2 What are Gift Cards?
2.13.3 Gift Card Benefits
2.13.4 Customized Design and Online Gift Cards
2.13.5 Terms and Conditions

2.13.1 Introduction

The creation of the gift card is one of the most revolutionary inventions for retailers in the last few years. Total annual sales of gift cards alone in 2007 are expected to exceed $35 billion. The idea of redeemable gift cards was a surprising breakthrough success because it was just so simple. With the advent of the Internet and the proliferation of soft goods, gift cards have matured to the online channel as well.

Gift cards' popularity is attributed to the large market of gift givers—specifically people that do not know what to buy or do not have the time to buy a "real gift." They are convenient purchases to make, and can be customizable if ordered through online channels.

For retailers gift cards are perfect product to offer because it is a way to get customers to purchase something of high value relative to the cost of fabricating the card. The return on investment for gift cards is incredibly high. And they are also a great option for customers because they are making a purchase for someone knowing they will not have to return it. It is a win-win situation for all three parties, the retailer, the customer, and the recipient.

> **The ROI for gift cards is incredibly high given that the market for gift givers is increasingly large.**

2.13.2 What are Gift Cards?

Gift cards are pre-purchased monetary valued options that retailers provide to customers who, typically, are purchasing it for another person. Think of it as a smaller, more portable gift certificate. The customer can choose from a pre-determined value set by the retailer, usually ranging from $10 to $200, however there is no limit to how many cards you can purchase.

eCommerce Best Practices

They are alternatives to browsing the site in search for a suitable gift, whether it be clothing, shoes, electronics, etc. There are three kinds of gift cards available to customers. One used exclusively to a retail store, one that can be used at a variety of stores, and one that is available and distributable online.

Gift cards available online can work two ways. One is that customers order an actual, physical card that will be sent to customers or a person of their choosing. The other way in which a gift card can be sent is via email, meaning it acts more like an e-card: an electronic version of a gift card that can be redeemed online as well. The purchaser can email it to the recipient as soon as the purchase has been complete. The benefit of having an e-gift card is that the turn around is immediate and sometimes can be used the same day as it was purchased. For anyone that is strapped for time, this is a great time saver as recipients do not have to wait for standard mail delivery.

2.13.3 Gift Card Benefits

Besides being a smart solution for a confused gift giver, gift cards can also increase sales. They are more efficient than paper certificates and more customers are going into retail stores or online looking specifically for gift cards. In addition, when displayed prominently, gift cards have been proven to be bought impulsively, boosting the retailer's bottom line significantly.

Gift Cards vs. Gift Certificates

Gift certificates are a thing of the past. Though they act in the same way as gift cards, the benefits of changing from a paper system to a card system is astounding.

One major reason being that a card is smaller and more transportable. It can conveniently fit into a wallet credit card slot. That might not seem like much because paper can fold up into a smaller size, but in the days where receipts and notes are thrown around into purses or wallets, those certificates have a greater chance of being forgettable or even unintentionally disposable. Because gift cards are about the same weight and proportions of a credit card you are constantly reminded of them every time you reach for your plastic.

Another reason why cards are more beneficial is because they are capable of being "reloadable." This means that once the funds on the card are depleted, the user can "reload" more money back into it. In a sense, it acts like an exclusive debit card for that retailer or set of retailers, so that it only uses the amount available on the card's account but allows you to put more funds into it as you please.

Drives Sales

Gift cards are successful at increasing sales because when people actually use the cards they tend to spend a greater amount than what is on the card. If they use a $50 gift card towards a $75 purchase, their reasoning for surpassing the card's amount is that they only had to spend $25 out of their pocket for $75 worth of product(s). Even though the purchaser is buying a specific dollar amount for the gift card, gift cards actually motivate people to spend more than the amount on it.

> Gift Cards motivate people to spend more than the amount on the card because less is spent out of their own pocket, encouraging a larger purchase.

Because it has been so popular, customers are astutely aware of retailers offering a gift card option, and are now visiting the store and/or site with the sole intention of purchasing one. Since they are purchasing a gift card for somebody else, curiosity often gets the better of them and they naturally browse around the site or store for themselves. Upon checkout, they buy the gift card in addition to the items they found while browsing.

In retail stores, gift cards are typically displayed at the counter, right next to the register. They often serve to remind people that the need to buy someone a gift, and subsequently buying a gift card right there on the spot. This is called an impulse buy—a purchase that was not planned, but you cannot resist picking up. It is like the candy and magazine section at the grocery store checkout aisles where all of a sudden you find yourself needing that chocolate bar. Impulse buys can drive up sales significantly because everyone at one point in their life has the urge or need to buy a gift on the fly, and gift cards remove the effort in searching for one. Consider placing an e-gift card option during the checkout process, shopping cart or wish list page—as to mimic the success of gift cards in the retail store setting.

2.13.4   Customized Design and Online Gift Cards

As mentioned earlier, most gift cards available online can either be exactly like the ones for sale at the retail store and mailed to the recipient, or can be an e-gift card that is emailed to the recipient. If you choose to purchase a physical card that will be mailed to the recipient, most retailers give you the option of customizing the design on the card. Some retail stores have a few pre-selected designs to choose from at the counter, but the online store gives you a wider

eCommerce Best Practices

selection of colors or patterns. Not only that, but it also allows you to write a custom message to the recipient if you want to add a personal touch, and some even let you enter a dollar amount outside the typical set prices.

How an e-gift card works is that the recipient is sent an email within 24 hours of purchase with the personal message and the dollar amount available to use. A gift card number is also sent, typically designed to be redeemable at the online store and, depending on the company, their retail store as well (a bar code is usually attached to the email if redeemable outside of the online store). Terms and conditions vary from company to company, so if you have or are developing a gift card system keep in mind the specifications and stipulations you want to employ.

> E-gift cards are time savers for most gift givers in need of a quick buy. They are more personalized and convenient.

2.13.5 Terms and Conditions

Rules and regulations for the use of gift cards are different for every company. However, most companies have common terms when customers buy gift cards or e-gift cards. Some being:

- Gift cards are not taxable; however, the purchases made with them will be taxed appropriately.
- Coupons cannot be used to buy gift cards.
- If the dollar amount purchased with your gift card exceeds the amount available on your card, you are responsible for paying the balance.
- Re-selling the gift card is prohibited.
- Any unused amount available after using the gift card can still be used towards your next purchase.

Gift cards and e-gift cards are great options to give your customers. Even though the amount set on the gift card is fixed, to a certain extent, people still tend to exceed the amount given to them. This, along with impulse buys can boost sales up significantly.

Paper gift certificates are things of the past. Funds cannot be "reloaded" into them, they are big in proportion but small in weight making them easily forgettable, and they cannot be custom designed or personalized. Gift cards are proven to be more effective and popular for purchases.

eCommerce Best Practices

Gift cards are a good way to draw business to your store or website because it relieves the pressure of finding a perfect gift for someone and it gives the recipients the freedom to purchase their own perfect gift(s).

eCommerce Best Practices

2.14 Customer Loyalty Programs

2.14.1   Introduction
2.14.2   Common Strategies for Loyalty Programs
2.14.3   Marketing Strategies
2.14.4   Balance of Rewards and Loyalty
2.14.5   Measure Customer Loyalty Required
2.14.6   Identifying and What to Collect

2.14.1   Introduction

When people think of loyalty programs, they often think of airlines and frequent flier miles. True, this is an excellent reference to a type of loyalty program as it encourages repeat purchases with a particular brand, but there are a variety of programs out there that utilize eCommerce resources to sustain them.

Loyalty programs increase customer sales and retention values. They also impact a lot of other business issues like customer acquisitions, winning back customers who have not made a purchase in a long time, and reducing advertising costs. Customers are loyal because in their mind your product is of high quality, brand image is strong, user experience is satisfying, and there is personalized attention to detail for products and user needs—responsive customer service.

> **Loyalty programs should have a balance of desire and affordability to attract customers.**

A strategically designed loyalty program aids customer retention. It entices customers to come back and shop to acquire more reward points. It is also important to point out that there are three main factors which govern the success of a loyalty program—one is timing (when the loyalty program is initiated), second is criteria (when the customer would get the reward) and lastly, the reinforcement rate (which is how often the rewards come in your loyalty program).

Let's look at some common strategies used in loyalty programs to retain customer loyalty.

eCommerce Best Practices

2.14.2 Common Strategies for Loyalty Programs

There are several factors which govern the success or failure of any loyalty program. Sometimes not having an accurate target will translate to not gaining true customers. This
might result in defecting patterns in the customer lifecycle and so to overcome this, you need to make sure that the loyalty programs are affordable, multi-usable across different brand outlets of the same retailer and give unique benefits like free shipping. Some common strategies are:

- Reward every time. Continuous reinforcement strategies work to create loyalty programs that rewards customers every time they make a purchase, like free shipping or special discounts everyday, once a week or any another time-bound date.

> **Measure and reward customers based on types of purchase, frequency of visits, and the amount spent.**

- Fixed and variable rewards. Sometimes retailers reward customers who have shopped more than 5 times during the same week, or month. A different way of approaching this is that customers get an option to win either of the above options, which is similar to a lottery.

- Interval rewards. Retailers sometimes choose to reward the customers who shop within a specific time period in the month. This unassuming approach is easy to manage as it simplifies the criteria of the rewards system and keeps customers anticipating whether or not they will reap a reward for their purchase. Unexpected rewards are often the most memorable.

2.14.3 Marketing Strategies

Most of the retailers deploy loyalty programs that are exclusive to their brand, although some are partnered with other companies who are partners to the main brand. Marketers use this to leverage a lot of successive factors to ensure the loyalty program gets more customer retention.

Partnerships

In the market there are un-partnered and partnered loyalty programs and retailers get an added advantage with partnerships within the same brand umbrella. Some notable plus points are that customers can carry fewer cards and earn points more quickly.

Retailers sell their cross-sell and up-sell products quicker with concentrated promotions, development time and savings in costs. There is also a greater penetration by members and non-members due to brand assortments and rewards benefits.

Customer Loyalty is Important

For each and every loyalty program, the customer is the primary, most important factor in your successes and/or failures. So retailers should try to leverage all their loyalty program strategies to retain, or acquire new customers.

> **Customer loyalty is important for retaining existing customers and winning back dissatisfied ones.**

Customer loyalty does not come across with a single purchase or promotion. It is the lifetime value of each customer that counts and adds to retention. Marketers require absolute measure of customers—what are the interests being served, whether they are purchasing products from a single source or service within the retail site—all of these add up to create a well-rounded loyalty program suited to the individual customer.

Though loyalty implies a long-term relationship, you should understand that consumer behavior is conditional. You should address solutions that target existing loyal customers to meet their demands and reward them. Retailers should listen to their most valuable customers.

Branding and Loyalty

Customers will be loyal to your brand so long that your brand is loyal to them. Should you try forcing the brand? Absolutely not. Loyalty can only be built upon something with which the people can be loyal to. Branding is important and marketers should strategize methods where loyalty programs are brand oriented with a balance of product information.

2.14.4 Balance of Rewards and Loyalty

Some of the key reasons for shoppers to join a loyalty program are being rewarded for products and services they frequently use, shoppers get greater discounts on products and services if loyalty programs are used, and special member perks which most retailers allow for frequent shoppers or bulk shoppers.

Do not be too stingy with rewards because and those shoppers who get big rewards are more likely to continue their shopping patterns than those who get fewer rewards. This means that rewards have to be lucrative to emphasize a desire to change the behavior of the shopper. Retailers have to balance desirability and affordability to attract customers which is often tricky.

> Never let your best customers feel that their privileges are taken away.

So how do you tackle customers who are loyal but do not shop as much and customers who are loyal and also shop all the time? If you reward them both then there is no scale that balances the difference. Loyal customers who shop more than other loyal shoppers should be rewarded more in terms of their purchases, time spent and the variety they had bought.

2.14.5 Measure Customer Loyalty Required

Building customer loyalty does increase profits. Marketers leverage CRM, customer databases, warehouses as positive development systems that provide data to improve business intelligence and decision making. Customer data is vital as you need to understand each customer personally—their shopping behavior, their needs and desired rewards. The customers you know the best are the ones that generate the most profit.

Retailers need to build their own customer lifetime model and look at alternatives to direct investments like mass marketing, service and quality management coupled with relationship management and CRM to understand and maximize customer profitability during their lifecycle with the brand and its products.

2.14.6 Identifying and What to Collect

Though there are many ways to track loyal customers in order to reward them later, retailers should strategize methods that make the steps and results more quantifiable and identifiable. Whether the retailer makes the reward a simple discount, or points that get converted to a gift from the catalog or other services, you just want to make sure the customer is responsive to it or else it is a mute point. It is practical to implement a customer relationship management system (CRM) for customer data collection, analysis and use.

Loyalty programs give you a way to establish a long-term relationship with specific customers and tie their user profile to their individual buying patterns, to give them more personalized rewards. There is no limit to the amount of data you can and should collect. The better you understand each and every customer, their demographics, preferences and lifestyles, transaction and return histories, the better you can attain their loyalty.

> **Customer data collection is important to target and segment your customers for loyalty program rewards. Determining exactly what to collect is more important than collecting everything.**

eCommerce Best Practices

2.15 Wish Lists

2.15.1 Introduction
2.15.2 What is a Wish List?
2.15.3 Wish Lists Benefits
2.15.4 Wish Lists vs. Gift Registries

2.15.1 Introduction

Shopping cart abandonment causes frustration for many eCommerce retailers. Right when you think the customer is ready to checkout, he or she leaves before the transaction has been completed. In the beginning of online shopping, items placed in shopping carts began to serve two different purposes. While retailers thought items placed were on their way to checkout, customers were placing them there to mark the products they were considering to purchase. The customer is merely using the shopping cart as a holding station for their products, but did not guarantee checkout.

This is still in practice today, but why do people place items in the shopping cart with no intention of buying them? Because it provides a means of convenient product reference. Wish lists are answers to this problem. Once shopping carts are abandoned, the information is not always stored for the next time the customer visits the site. Putting items in a wish list allows customers to re-visit the products they were or are interested in.

A wish list is a deliberate product reference area that allows potential buyers to organize the products they are considering to purchasing. Though it may seem hard to break the habit of customers using the shopping cart as their reference area of choice, as they become more familiar with the site and the ease of adding items to the wish lists, the more comfortable they will be use it to their advantage.

> **Wish Lists reduce the risk of Shopping Cart abandonment.**

2.15.2 What is a Wish List?

Wish lists allow customers to save products to a list for later purchase or to send to a friend, co-worker or family member to purchase for them. Customer can store items here without placing them in the shopping cart, so that they do

eCommerce Best Practices

not feel pressured or obligated to buy them. Wish lists are also a good way to bookmark a lot of items and compare them in one central location on the site.

2.15.3 Wish Lists Benefits

In addition to the organizational benefits to customers, retailers are reaping the benefits of wish lists as well. Retailers are finding a reduction in shopping cart abandonment, more repeat customers, an increase in sales, and a means to acquire valuable marketing information.

Engages Customers

Once potential customers arrive or are driven to a site, it is important that the website further engage them to navigate through the site with the intention of them meeting their goals, i.e. purchasing their desired products. This is done by creating a site that has an appealing design with meaningful images, implements targeted promotions that optimize purchase potential, and is easy to navigate from page to page.

Yet, even with all of this, distractions are inevitable and people simply may not have enough time to purchase items the first time they visit the site. In the days of power lunches, soccer practice, school, work, meetings and remembering to breathe, it could take multiple site visits in order for an order to be placed.

Unlike shopping carts, wish lists store recently viewed or saved items so when customers re-visit the site they can pick up where they left off during their previous purchasing process. This removes the need to re-search for certain products and reverts people back to the mindset they had when they first created the wish list.

Drives Sales

A consumer is less likely to forget about products when they create lists for themselves. It serves as an intentional reminder of products they are considering to buy, but may not have the time or means of buying them at the moment. Likened to impulse items at a checkout counter, wish lists are capable of boosting sales in an online setting.

A wish list immediately establishes a relationship between a consumer and a website, whereas without it there is no lingering connection post site visit, which leaves more room for a consumer to stray. And since there are

eCommerce Best Practices

so many offline and online competitors in the market, the consumer-website relationship is key for driving sales and having a slight edge over the competition.

> **Wish Lists can boost sales by creating an immediate relationship between the consumer and website, yielding more potential to boost sales.**

Word-of-mouth, and word-of-mouse in this case, plays a significant role in increasing sales and directing people to the website. If a consumer does not have the means to purchase the items on his or her wish list, he or she can easily inform a friend or family member about the wish list, in which they can make the purchase as a gift. There are many different occurrences like a big pay-check or a birthday that can spark a purchase, and even if it is three or four months from the time a wish list was created, those items placed there would be readily available to accommodate any occurrence with purchasing potential.

Rich Data

Perhaps more valuable than reminding people of items in their wish lists and boosting sales, the marketing information obtained directly from consumers' creating wish lists serves a bigger purpose. Retailers can set-up a data feed that enables them to look inside each consumer's wish list and wish list history. From this alone, you would know what types of products every visitor is interested in and the budget he or she is willing to spend. In the future, targeted promotional offers or ads can be filtered to suit the interests of that specific consumer. The type of targeted marketing effort enhances the potential of repeat buyers, website loyalty and an increase in sales.

> **Wish Lists provide valuable, targeted marketing data like specific products of interest and budget range, which can lead to a longer and more loyal relationship in the long-run.**

2.15.4 Wish Lists vs. Gift Registries

Most eCommerce retailers give the customers the option to set up the lists as private or public so not just anyone can view all aspects of a private wish list. Gift registries are more open to the public. They allow someone to basically create a wish list of all items he or she wants and then leaves it to other people to purchase the items as gifts. Gift registries are searchable by others and are typically used for weddings, baby showers, or other big events in someone's

life. The product information is stored and anyone who has received an invitation to the party will be able to view the gift registry and purchase items from the list.

A wish list might be set up as a private list, in which case, it is similar to a gift registry except that it is only accessible to the person who created it. Customers can access their wish lists and purchase items from it at any time.

The creation of wish lists and the gradual adoption of them into more and more eCommerce sites have made strides in reducing shopping cart abandonment and increasing conversion rates. A major benefit to them is that they are able to store items placed there, allowing users to quick reference items from the past and make purchases several months after the fact.

eCommerce Best Practices

2.16 Order Management

2.16.1 Introduction
2.16.2 Value of Order Management
2.16.3 Multi-Channel and Inventory Control for Order Management
2.16.4 Steps to Select Order Management Systems

2.16.1 Introduction

An eCommerce retailer would not be able to thrive without an effective order management system to handle functions as diverse as order entry, sales analysis, inventory planning and accounting. Today we have web-enabled order management and fulfillment systems that allocate inventory to order, process payments, manage the shipping schedules and generate the confirmation/packing documents. These are a few of the system tools that must work cohesively in order to produce a fully funtional site.

> **Drive customer satisfaction by timely fulfillment and flexibility.**

As retailers grow from multi-channel selling to broader geographical regions, inventory levels increase based on demand. When a retail site increases the number of sales channels across online, catalog, retail stores and business partners, then the coordination of the order fulfillment process from capture to fulfillment becomes a complex process to manage.

Customizing your order management system may be necessary if you have a specialized ordering process. If your site has specific requirements that customers need to adhere to in order to make a purchase, i.e. geographic location, age, affiliations, etc., your order management system needs to developed in support of those specifications.

2.16.2 Value of Order Management

Retailers are looking for cost-effective ways to ensure that what the customer ordered is delivered on time and at the right location. Customer deliverables are based on the capacity of the retailer's fulfillment network or on the capabilities of their internal distribution resources.

A typical hard goods order process is as follows: a customer order is received, the order management system queries the distribution centers to see whether the merchandise is available, and packages and sends it out based on the selected shipment type. If the merchandise is not available at the distribution center and there are retail locations, it would check the inventory at stores closest to the delivery location of the customer. Once the store is identified, a pick ticket is issued and it becomes the store's responsibility to determine whether the product is available or not, and then confirm when it has been shipped to the customer or is awaiting in-store pick-up (should the customer opt for that method of fulfillment). Of course, it is important to have the most up-to-date inventory status of every product to make sure that the availability is accurate.

Effective order management can optimize store order replenishment in retail stores that sell fast moving consumer products by driving replenishment from the supply network, which can be a local retail distribution center, a local vendor, or a packaging plant. Some of the unique elements of effective order management systems are:

- Regardless of channel, it should provide a similar, single interface for the customer and customer service representative as to minimize confusion during assistance.
- It should clearly display the real-time status of the customer's orders and is flexible to change orders when needed.
- Real-time inventory view is available across locations, including suppliers or logistic partners.
- Supports inventory checks and provides receipts, production schedules and also order and delivery commitments. It should be effective in dynamic sourcing of inventory across all locations to fulfill orders based on business objectives like revenue, lowest delivered costs, and customer satisfaction.
- Shipping and handling costs can be lowered by providing multiple options. Also, it facilitates trans-shipments and inventory deployment.
- Provides enterprise and supply chain-wide visibility of order status and fulfillment operations across your entire retail network.
- Drives customer satisfaction by increasing fulfillment speed and flexibility for the customer. Retail customers can choose to pick their order from a retail store (if available) or having it home delivered.
- Customer or sales representative alerts related to any order status change allow for convenient notifications, and they also minimize the deviation which might have occurred due to a lapse in communication.

2.16.3 Multi-Channel and Inventory Control for Order Management

Multi-Channels can improve profitability and, even though it involves a more complex order management system, retailers are finding it more advantageous to offer different selling points. Retailers who sell across the Internet, catalog, and/or retail store are more easily accessible in terms of making product purchases, thus, more likely to be more profitable than retailers selling through one channel. In fact, retailers are finding that many customers start with looking at one channel and end up purchasing with another. For example, someone may receive a catalog in the mail and then make a purchase online based on what they saw in the catalog. Managing the different channel operations certainly adds complexity to product management and order fulfillment.

If you are considering expanding to multi-channels, put into consideration the amount of inventory to prepare for. Logically, if the number of orders increases then you must have the inventory to supply them. Retailers that manage multiple distribution centers and that hold the wrong amount of inventory in the wrong locations end up with customer service failures. The influence of customer service on future transactions is astounding, so getting the orders fulfilled is extremely important especially if you are beginning your multi-channel endeavors.

> **Multi-channel purchases are more profitable when customers experience effective order management and fulfillment processes.**

Many retailers lack the technological support to determine the most cost effective way to fulfill an order, including triggering inventory execution activities across multiple stocking locations and across multiple partners. Retailers should assign a rule with each customer to be assigned a warehouse that is most geographically beneficial, but also provide a system that is able to balance and deliver orders incase of shortage of stocks. This is best so that customer orders are not delayed. Distribution and fulfillment centers are most beneficial to you when it is located around the Corn Belt of America—typically in the Midwestern states, i.e. Kentucky, Iowa, Wisconsin, etc. The reason being, that shipping costs from those areas are relatively low, which gives you more room to offer a free shipping promotion, and the ground delivery time is around one to two days for many locations.

eCommerce Best Practices

Inventory should be able to be accessed from different product sources—stores, multiple distribution centers, suppliers and logistic partners. Suppliers and logistic service providers are essential parts of the order fulfillment cycle and retailers should extend their processes to improve performance and customer service.

2.16.4 Steps to Selecting an Order Management System

Replacement or acquiring a new order management system is a daunting task. How do you resolve and make at the right decision? System project teams within a retail company are the best go-to option as to minimize evaluate the limitations of your time frame, which can vary depending on the complexity of the requirements. However, when it comes to evaluating potential solutions, there are some standard steps which you can follow:

> **Standards are lacking in most systems: relate their system objectives, operations and budgets to implement a standard direct-commerce order management system.**

System Objectives and Priorities

Managing increased capacity, improving customer service online and offline, business methods like continuity shipping, lowering processing costs, and inventory efficiency are some of the reasons for acquiring a new order management system. It is always important to prioritize the goals so that the processing is made easier.

Operations and Budgets

You need to spell out how the order management system will serve business requirements in a formal and comprehensive functional method. Put into consideration your needs for a cohesive flow of operations between different departments like warehouse, fulfillment, and customer service, etc.

RFP to Provide Analysis

Besides the results of the functional needs, the RFP should also provide requests for vendors to indicate if they can meet your requirement, length of time it would take to implement and also any support from out-of-the-box functions. Listing of all your hardware, networks, and explanation of

how your retail business works if there are any third party processors, are all equally important to be included in the RFP list.

System Options

Ensure that the RFP is given to more than one contender to give you a choice when deciding and evaluating the appropriate order management system. Depending on your scale, number of products and SKUs, you should make sure the system can meet your current and future needs.

Uniformity of Standards

Each order management system works differently and the lack of standards is a big issue to address. There are three aspects of standards: uniformity in functions, methods, and formats. Currently, there is no set standard in direct-commerce order management systems as virtually every system works in a different way. Whatever the difference is, the order management system should cater to the retailer business requirements based on system objectives and operations.

Vendor Investigation

Retailers can also inquire from the vendors an initial request of information which details technical data about each order management system being provided. Scheduling demos are another option to get a first hand feel of the product and decide whether it meets your business requirements. The demo process should be able to help you to narrow down the vendor listing to a few probable options.

Alternative Reviews

Retailers and marketers need to know that there is no perfect order management system and vendor. Once the vendor list is short listed, you should step back and review whether all bases are covered: call center, eCommerce, warehouse, fulfillment, reporting, and accounting.

eCommerce Best Practices

service

ecommerce
bestpractices

SERVICE

The relationship between a business and its customers does not stop once the sale has been made. Any good retailer knows the importance of customer satisfaction before, during, and after a purchase, as every point of contact influences the overall impression of a brand.

Customer service, inventory management, and product purchasing knowledge, each contribute to the service or experience of the online customer. The failure of any one of these factors to exceed customer expectations, risks portraying the organization as a whole in a negative light. It also places future growth expectations in jeopardy. Customers will remember any unpleasant experience, and may simply move on to a competing retailer.

Creating a lasting, positive feeling for online shoppers during the entire end-to-end purchasing experience should be the primary goal for an organization's eCommerce initiative. This section will detail some of the successful service components of the site and address pitfalls to be aware of:

- Developing your Online Customer Support
- Efficient Claims / Returns Procedures
- The benefits of Channel/Partner Extranets
- The advantages of creating a Forum
- Warranty and Maintenance policies
- What to consider for your Online Manual
- The importance of Call Center Support

eCommerce Best Practices

3.1 Online Customer Support

3.1.1 Introduction
3.1.2 Attributes and Features
3.1.3 Technologies and Issues
3.1.4 Recommendations

3.1.1 Introduction

More and more, today's customer's are demainding immediate crisis resolution. Whether that is as simple as forgetting a password or as complex as device-driver downloads, online support is critical to maintaining customer satisfaction. One popular tool is using a customer-focused knowledgebase as the organized collection of shared content which allows for a logical reasoning process or inference engine to derive context based information from it. Most implemented knowledgebase solutions are either application-driven or dependent upon user input for sustainability. While the former provides cookie-cutter solutions to the user, the later often contains more greater content that may or may not be applicable to the user's problem.

> **Online knowledgebases are becoming an important commodity and promising service business.**

With widespread adoption and use of Internet technologies in all aspects of organizational enterprise activities, an online knowledgebase solution has become a mandatory tool for creating a world-class, best-of-breed service business. This has been brought about by improvements in knowledge storage and presentation technologies, as well as advances in automated reasoning techniques.

3.1.2 Attributes and Features

The most important objective of any online knowledgebase solution is to provide contextual information about systems, processes, tasks, products, services and procedures to users as needed. From a quality perspective, it is critical that information supplied to a knowledgebase solution be

> **More advanced knowledgebase should be self-learning and self-improving.**

complete, timely, structured and well formatted.

Various forms of information exchange and content representation techniques exist. Some of the most popular include:

- Intelligent systems with structured knowledge, facilities for advanced state space search and an inference engine which is capable of learning from experience.
- Meta language for rules and logic that is represented in a standard way.
- Semantic web for representation of web that enables interpretation of documents by software agents.
- Guides, tutorials.
- Wikipedia pages (or idea banks)
- Query center that allows user to post questions or topic and get explanations or answers.
- Enhanced search facility.
- A learning inference engine.
- Flexible, dynamic provision of answers.
- Contextualized and organized knowledge.
- Ease of authoring, intuitive workflow.
- Enhanced reporting tools.
- Core process that tracks the usage pattern and drives the system toward improvement.

> **Wikis and idea banks are only the beginning of building a knowledgebase, but may meet all of an organization's requirements.**

As far as content output is concerned, the results of a well-structured online knowledgebase should include the following characteristics:

- Support solutions at the time of need.
- Elimination of repeat efforts.
- Greater collaboration.
- Dynamic knowledgebase.
- Quicker problem resolution.
- Ease of document updates.
- Automated response capabilities.

> **Collaboration is critical. It is essential to consider the opinions of all stakeholders in developing a knowledgebase.**

eCommerce Best Practices

3.1.3 Technologies and Issues

From a client delivery standpoint, numerous standards and technologies have supported the proliferation of online knowledgebase solutions. Some of them are listed below with their most outstanding feature.

XML (Extensible Markup Language)
XML is a standard tool for defining and representing data in an interoperable and transferrable manner. XML is the best-of-breed solution for integrating with third party disparate applications.

XML Schema
An XML Schema is a framework that defines and limits the structure and content of XML documents.

RDF (Resource Description Framework)
RDF is a model for making references to objects and their relationship. This simple data model and ability to adapt to disparate data references makes it invaluable for online knowledgebase solutions.

Semantic Web
Semantic web is an important principle that ensures that websites can be processed by applications and properly parsed and classified at every level. The central principles of semantic web are formal specifications for representing documents and ability of software agents to get knowledge from those documents.

With the possibility of software agents to interpret the semantic web, it then becomes possible to automatically perform search and analysis based on web documents. Semantic publishing is a potential service sector that will benefit from semantic web.

Notwithstanding its great potential, semantic web has some issues, most notably, it is more time consuming to create and publish documents. For any document created and published, it would essentially require one version for human viewing and another for machine interpretation. This creates additional efforts and, correspondingly, more time for the documentation process. This is being addressed in development so that a uniform format can be viewed by humans and used by machines alike.

eCommerce Best Practices

Intelligent Systems

Intelligent systems encompass the structure and organization of a knowledgebase with the objective of enabling a machine to infer contextual knowledge or intelligence from it. It uses advanced state space search and inference techniques based on highly efficient knowledge modeling.

3.1.4 Recommendations

Online knowledgebases are rapidly evolving, but are also being adopted by more and more business sectors. Publishing, organizing and making web documents searchable is only the beginning of online knowledgebase capabilities.

Wikis have done much to popularize the knowledgebase arena, so customers already have an enhanced expectation of collaboration and ease of use.

> **It is important to weigh the cost and efforts incurred in building online knowledgebase in relation to the requirements and intended outputs.**

Building the semantic web and developing intelligent systems are the next great evolutionary steps in knowledgebase development.

There are significant costs and efforts associated with building semantic web and intelligent systems, but the returns may be worth this investment.

As the online knowledgebase sector matures, and standards are put into place, the effort and cost incurred to build an online knowledgebase decreases.

eCommerce Best Practices

3.2 Product Claims & Returns

- 3.2.1 Introduction
- 3.2.2 Reasons for Returns
- 3.2.3 Solutions Provided by Software
- 3.2.4 How to Reduce Returns

3.2.1 Introduction

Retailers know that when a customer is not satisfied with a purchase and makes a return, the transaction should be as quick and painless as possible. Returns also play a key role in your inventory management — the ability to balance return rates within inventory levels is important to prevent overstock and backorders. Even though the customer is returning a product, an efficient return process is an opportunity to retain that valuable customer relationship. An efficient return process with therefore build customer loyalty and enhance your inventory management abilities.

Though most of the fulfillment and catalog management systems have some ability to report return statistics, you should be able to analyze returns by vendor, item and quantity as a percentage against the units that were shipped. You have a better chance of reducing future returns if you know why products were returned in the past.

3.2.2 Reasons for Returns

Due to the popularity and shear volume of transactions the eCommerce world has been able to drive, there has also been an increase in return rates. Retailers and marketers are often left wondering what went wrong after the customer received the product. Even though some of the factors are simply beyond the control of any retailer, understanding as much as possible about the return is critical.

> **Communicate with customers about the return processes through information on pages, on receipts, and within a packaged delivery product.**

A product is usually returned for one of five reasons:
- Poor product quality
- Slow delivery or shipping

Part III: Service

- Wrong product shipped
- Product described incorrectly
- Customer choice

Customer choice is beyond any retailer's control, but the other four reasons can be partially mitigated by improving product management and operational efficiencies.

Picking and Choosing

Today, customers are using their own homes as "fitting rooms" for the products they buy online. Customers may purchase a variety of products in order to "test them out," and after they find the right color, size, and/or fit, they keep the ones they like and return the rest. Some product vendor's offer free shipping for returns, and some consumers will take advantage of this by purchasing more than they intend to keep, and returning the ones they do not like. An enhanced, more interactive user experience can reduce this behavior, giving the customer gets a better grasp of what they are purchasing before it arrives.

Unfulfilled Expectations

This is sometimes ignored as marketers and retailers should know that one of the reasons for increased return rates can also be the merchandise itself. What looks good in catalogs or online may not be what the consumer had in mind—color and size expectations may fluctuate from retailer to retailer.

3.2.3 Solutions Provided by Software

Retailers who have multiple distribution centers or stores, have to make sure that properly managed returns can make up for inventory shortcomings at various locations. There are vendors that provide return software that facilitates the redistribution of returned merchandise. These solutions are capable of handling all phases of the return process — everything from processing the return requests, to sending the credit back, to creating a receipt for the exchange. These can also be routed to the catalog inventory management system which returns the merchandise to the warehouse or retail store that needs it the most.

3.2.4 How to Reduce Returns

There are still some reasons for returns that are within your control. Unclear copy or graphics that are hard to associate from the catalog to the web can be reshot or otherwise improved. A lack of details or factual errors in a description can also result in poor product expectation management.

Successful retailers minimize returns by measuring and analyzing the particular causes of returns, calculating the actual cost of returns and creating a plan to reduce the costs and maintain higher levels of customer services.

Identify Reasons for Returns

While some reasons for returns are virtually unknowable, there is much to be gained by carefully analyzing return data and determining or estimating likely reasons for the failed transaction. Many retailers find it useful to identify key merchandise categories, and then determine which are more likely to generate returns. Return reports may include returns based on these categories, as well as promotions, vendors, items and the stated reason for a return. This is particularly important when retailers are faced with high-return categories. The data provides a transparent view that shows why people made returns, whether it is from poor design, damaged goods, or shipping/packing errors. Also misleading presentations and pricing on promotional materials can cause customer disillusionment and lead to returns. Analyzing compiled data enables you to see which departments or procedures within the retail organization are contributing to costly returns.

> Data collected during the returns process can be used to calculate costs of returns which in turn aids the action plans.

Calculate Costs of Returns

Once the reasons for the returns of merchandise are detailed, the next step is to calculate the cost of each step in the returns process. This can include cost of materials needed for packing and repackaging, the number of man-hours involved in receiving and processing returns, staff involvement for customer credit cards, inventory management files, etc. When the returned product is damaged during shipping, retailers usually sell them at discounted prices that translate to still higher cost. The most vital, yet intangible cost factor is the state of the customer's relationship and value when they are forced to return a purchase.

Reducing the Cost of Returns

Once the reasons for returns are discovered, an organization should work to prevent further returns for similar reasons. Priority should be given to the reasons that are the most costly, but the cost of solving the problem should play a factor as well. Fixing one high-cost problem quickly may have a more immediate return on investment that solving five minor issues, but if that one solution is prohibitively expensive, it may be best to resolve the smaller problems that represent the "low-hanging fruit" of your return issues.

> **Return processes should be quick as customers do not have the patience. Customers are more likely to shop with the same retail merchant again if the return process is convenient.**

Communicate your Returns Policy

Organizations that ensure customers are aware of their return policy before they make purchase are proactively reducing support calls and angry customers. Retailers should provide detailed return information on their site, in the catalog, and also at the store. Return policies should be as detailed as practical, but not overly complicated.

Return Process Should Be Quick

It always works in your favor if the return process is fast or even faster than the purchases. Since customers have less patience during the return process, retailers and sales teams should expedite the credit or exchange product.

Cross-channel Returns

Shoppers who have purchased items from one channel – such as a website, or storefront – should be given the capability to make a return through another retailer presence when possible. Some retailers accept returns at their stores even though the product was bought online. Retailers need to handle this option carefully as online catalog and stores might have different product lines, which would require special handling for a unique return.

> **Customers expect cross channel returns to work just a quickly as standard returns, in spite of the added complexity**

eCommerce Best Practices

Smart Return Processes

Reverse logistics is very important to retain customers, and it describes the process for handling all returned products, once the customer decides they do not want the particular product. Pre-printed return labels and smart labels enable customers to adhere to the organization's return policy, and makes returns as easy as possible. Labels can also be generated online, further speeding up the returns process. Bar codes or other machine-readable labels are ideal to quickly the product so that it can be returned to inventory or disposed quickly and efficiently. RFID tags are also being considered for speedy processing of returns, but this method is currently best described as experimental.

Returns are an unavoidable part of merchandising for all retailers, but if handled properly, they can be yet another opportunity to enhance the relationship between customer and merchant.

3.3 Channel / Partner Communication

3.3.1 Introduction
3.3.2 Extranet and Its Benefits
3.3.3 Channel and Partnerships

3.3.1 Introduction

When business sales slow down and the market becomes even more competitive, retailers often look to form partnerships with other parties. Partnerships and strategic alliances become more appealing during these unstable times. When considering partnerships, you should approach the alliance with care as it could be disastrous for all parties involved. For example, there have been instances where manufacturers alienate their distributing partners, which often attributes to the failure of the partnership programs.

For B2B companies, channel partners provide a selling model and an integral part to the success of the company's profit or loss. The use of an secure shared network — called an Extranet — can be more significant than the affiliate marketing program used as a medium of communication and interaction between company and partner. An Extranet cannot compensate for a partner program that is flawed or unstable, but it can have a major positive impact on solid relationships formed with any type of partner.

3.3.2 Extranet and Its Benefits

An Extranet is essentially a private web site or network unseen by the general public. Only members of the partnership that are given access have the ability to view the information on its pages. One of the best uses of this technology is enhanced customer support. Similar to offering a 'client only' Extranet where all types of communications like online support, product updates and releases, delivery etc. are provided, retailers can pair the same features for their partners. Each partner has their own private space to communicate solutions to problems, present and resolve issues with the client, and interact with marketing or IT team members for specific queries.

> **Use an Extranet to conduct actual business with partners, not just as a marketing support medium.**

eCommerce Best Practices

3.3.3 Channel and Partnerships

Some of the ways a retailer can combine principles of partnering with Internet marketing are:

Partner Service Extranet

Retailers can create a web site that is exclusive for their partner or partners. The web site can be a part of the retailer's existing site with an exclusive private access area that only the partners can view. Another option is to have a private extranet that uses a separate URL hidden from public view. In both cases the main objective is the same—to provide a site that services the partners.

> **Partner extranets may be an exclusive domain, or a secure area within an existing site.**

Promote Partners in Special Slots

Retailers who have special partner relationships should promote the same across their corporate site. The common method is creating a special section within the eCommerce site dedicated to partnerships and partnership growth. It is intended to attract more new partners by highlighting current or new partnerships, benefits, news, and allowing partners, themselves, to describe the program.

Communications Link

Retailers should encourage their retail partners to communicate with partner employees. Better communication leads to greater cohesiveness between all parties involved.

Collaborate Online With Partners

Encourage partners to build relationships across the web and promote their information on eCommerce sites. Retailers also can have a "partner showcase" section within their site where partners can post information useful to your products and services.

> **Partner Showcase sections within retail sites allow partners to post useful information for marketers and shoppers.**

Retailers should understand that the strong value of a partner extranet is service. A full range of promotional and marketing services of the partners

should be offered through the partner Extranet. Paper-based systems can also be used to service the partner relationship through the Extranet. Product ordering, lead distribution, results tracking, invoicing, inventory tracking are some of the services that can be used. The Extranet should not only serve as a support medium, but also be used to conduct business with partners.

eCommerce Best Practices

3.4 Monitored Forums

3.4.1 Introduction
3.4.2 Forums Styles
3.4.3 Pros and Cons
3.4.4 Desired Features
3.4.5 Third Party tools

3.4.1 Introduction

A forum is an area on a website for holding discussions with large numbers of people united by a common interest. Each forum typically contains a number of sections, each of which contains threads of discussion. The number of major forum sections usually remains static, however the administrator can make additions if need be. New threads are posted on the forum by users, which generally hold no limitations to the number of threads they can post. One or more users may be given moderator privileges to make sure that the topics being posted are relevant to the intent of the forum and to make sure proper netiquette is being followed.

In the context of eCommerce websites, forums are focused on providing a facility whereby customers and users can post a request for help, support, or general questions about products, company policies, or promotions. The answer can be provided by other forum users or a company representative whose function is to provide timely answers to user queries.

> **Anonymity leads to interactivity. Forums give people a platform to express themselves or ask questions without confrontation.**

3.4.2 Pros and Cons

Using Forums can add depth to your site and make it a one stop shop for product queries, purchases and troubleshooting. However, there can be a number of drawbacks to adding forums to your website. Listed below are some of the advantages and disadvantages of forums.

Advantages

Customer Loyalty and Satisfaction: Forums can increase customer loyalty if it evolves into a site where customers can come for all product related information.

Interweaves sales opportunities into community features: Forums can provide an informal medium for interacting with potential customers and finding insight into customer needs and wants.

Drives more visits to the site: More visits to the site means more possibilities of users buying products, especially if promotions or cross-sells are integrated with forums.

> More people are driven to sites with forums, especially now that they are being associated with exclusive promotions.

Provides for informal product feedback: People often feel more open to discuss issues and provide feedback in a community of users rather than giving formal feedback to a customer service representative.

Can be a tip-off on viral trends: Forums can act as early warning systems for upcoming trends and provides a way to demand forecast when preparing your inventory.

Disadvantages

Investment in human capital: Forums would require one or many moderators to make sure that the rules of conduct are being followed by its members. Failure to do so might drive away people from the website as well as negatively effect customer loyalty.

Might require domain/product experts to monitor the site: If purpose of the forum is to provide support for products, it will require someone with some degree of knowledge about the product to answer queries.

One size does not fit all: Forums might not be meant for all websites. These are most suitable for providing tips, support and troubleshooting (such as cleaning products, electronics and machinery) as opposed to products such as clothing items.

3.4.3 Desired Features

There are a number of features you might want to consider before investing into one of the many types of forum software (either free or paid). Some of the items

eCommerce Best Practices

that should be considered before investing time and possibly money into integrating a forum into your website are listed below:

Database support: Typically, only those forums that support the same database as your main site should be invested in. This saves on procuring new resources with the necessary database expertise and allows for easier backend integration.

Calendar and Events: This feature would be relevant if you expect to host a lot of events for the participating community or want to allow customers to publicize their own events. Please note that this functionality might serve better at the main website if the event is targeted at all customers as opposed to those who only participate in the forums.

Image Attachment: This functionality would be important if attaching images serves the purpose of the forums such as sharing images of user customization such as a photograph of a custom home theatre system.

Programming Language: This would be a consideration if a third party forum package is being used and requires new custom functionality.

Sub-forums availability: This would be required if a hierarchy of topics/forums are to be created in the forums site.

Karma points: A user earns karma points for posting insightful or useful comments. An eCommerce site can encourage forum participation by making karma points redeemable for coupons or gift-cards.

> **Consider offering redeemable karma points, either through coupons or gift cards, for users posting insightful comments.**

ACLs: Access Control Lists (ACLs) could be utilized to giving/restricting access to certain groups of customers. An example of this would be preventing B2C customers from reading B2B forums.

The features mentioned below do not directly affect the workings of an eCommerce website, but are important functionalities that should be considered when investing in integrating, purchasing or implementing a forums solution.

Polls: Polls allow for explicit user feedback and would support the business objective of gauging user feedback. However, if the poll is geared towards all

users, instead of just forum participants, then it would be better purposed on the main site as a separate implementation or addition.

Moderation features: These features include moving, splitting and closing threads to better maintain the organization of forums and sub-forums. It may also entail deleting off-target or offensive messages or even whole conversations, as needed.

Notification (RSS, email): This functionality allows for users to opt into being notified of new posts or replies. They can be notified either by RSS or email. This would increase more traffic to the site if the user finds out in real time when someone replied to their posts.

Spam Control: This is to prevent "bots" from going into the forums and spamming it with ads and commercials.

Flood Control: This is an administrative setting similar to spam control, but prevents users from posting excessively.

Warnings, Suspension, Blacklist, and IP-blocking: This allows moderators to maintain decorum in the forums should any user try to become abusive.

Archiving: If the number of posts is expected to grow very large, the archiving feature would allow you to archive older posts. These would still be available but would be displayed at a slower speed than the current postings.

3.4.4 Third Party Tools

Below is a list of some prominent third-party forums available online. There are many more forums than the ones listed, but these are representative:

- Discussion Board - http://www.activedataonline.com.au
- FuseTalk - http://www.fusetalk.com/
- Gossamer Forum - http://www.gossamer-threads.com/scripts/gforum/
- Jive Forums - http://www.jivesoftware.com/products/forums/
- phpBB - http://www.phpbb.com/
- vBulletin - http://www.vbulletin.com/

3.5 Warranty and Maintenance Contracts

3.5.1 Introduction
3.5.2 Warranties
3.5.3 Maintenance Contracts
3.5.4 Maintenance Checklist

3.5.1 Introduction

Successful retailers maintain their reputation by relying upon efficient, cost-effective procedures to procure, store, distribute, and sell safe, high-quality products. Many of these products are manufactured and maintained by technologies requiring warranties and maintenance for ensuring consistently high levels of quality.

Organizations should clearly understand and maintain their legal terms for these grounds with their respective third party vendors and also protect their products during sale processes to customers. Warranties and maintenance also extend to the entire purchase cycle process, therefore every part of an eCommerce site is also included in this understanding. Retailers violate consumer protection rules regarding warranties, shipping claims and rebates at their peril.

> **Retailers should always inform customers about delays or other shipping problems, as these might invite legal implications if the information is not clear on the site.**

By Federal law, all products for sale require written warranties. Retailers should also ensure that their advertising is not misleading, and have their warranties or rebates included in each advertisement. They should clearly and prominently disclose the type of rebate (if one is offered), the process to redeem it, and the total price the consumer must pay at the time of purchase.

3.5.2 Warranties

Warranties are legal contracts between retailers and customers. They detail obligations which call for the replacement or repair of tangible personal property with either no additional charge, or a reduced charge for parts or labor. They also refer to the replacement or repair of tangible property based upon

eCommerce Best Practices

happenings of some unforeseen occurrence or event that renders a faulty product, within the warranty period. Some retailers also give extended warranties which require additional charges to the customer.

When a warranty is included in the retail selling price of the product being sold, the value of the warranty becomes a part of the selling price. Online products are no exception, and retailers should recognize how the value of warranties will fit into their selling price.

> **Warranties and maintenance information should be part of the customer education process.**

3.5.3 Maintenance Contracts

Retailers often provide maintenance contracts or agreements for customers or other parties associated with their products. These agreements, referred to as service contracts, require specific performance repairing, cleaning, altering, or improving of tangible personal property on a regular or irregular basis to ensure a product's continued satisfactory operation. Retailers should know that maintenance agreements are separate retail sales and are subject to retail sales tax. So, like the warranty terms, maintenance agreements also carry sales and use tax.

3.5.4 Maintenance Checklist

The maintenance of IT systems within the distribution center of a retail outlet can be incredibly complex. Applications like enterprise resource planning (ERP) systems, warehouse management systems (WMS), warehouse control system (WCS), and others like labor management system (LMS) and transportation management system (TMS) are just some of the integration systems that top the maintenance checklist. A problem with any of these applications could cause damage to the entire system.

To avoid such problems, it is imperative to understand how data is originally collected within the various applications. This understanding can uncover deficiencies in processes or controls or even the base applications. These deficiencies can then be avoided by providing additional controls within the process of analyzing the base data for business intelligence applications. Software updates and maintenance releases are a must for all retailer systems.

> **Retailers should review the upgrades and customization in detail, and consider testing as well.**

Different levels of evaluation are necessary during the maintenance period. Like at the operating system level, you need to check whether appropriate updates have been applied to ensure proper levels of security. At the application levels as well, upgrades should be installed and checked to improve performance and business setups. Lastly the network level, virus and spam protection needs to be checked and up to date. If you do not pay attention to these basic checks then they could end up using far less of their system's capacity. Even though users took in their product for maintenance, if you return that product in a better condition than what was originally purchased, a pro-active and thorough repair is more valuable than the availability of the maintenance service itself.

Maintenance checklist disruptions and warranty issues are is lessened today because of the popularity of SAAS (Software as a Service). All manner of systems like order management, contact center management, customer relationship packages are included. The benefit for retailers is that they do not have to make the capital investment in servers, software, IT infrastructure, and management responsibilities.

3.6 Online User Documentation & Manuals

    3.6.1   Introduction
    3.6.2   Need for online manuals
    3.6.3   Content distribution

3.6.1 Introduction

Online manuals provide an excellent opportunity for customers to become familiar with many aspects of a product before an actual purchase. No matter how targeted the email listings are or how inviting the product descriptions are, you should ensure that each product has a relevant online manual within a category that is easily identifiable to the common user.

Choosing the proper document format is an important part, and standardization is a key factor. Customers can easily become frustrated they are unable to view a particular product manual. Some retailers also provide direct links to manuals and other product details in emails. The effectiveness of these links are different for each retailer, so be sure to implement a discovery period to test it out.

3.6.2 Need for Online Manuals

Retailers should determine which of their products would benefit most from the availability of online or downloadable manuals. Generally, the more complex the operation of a particular product is, the more likely an online manual will help convince browsers click the "add to cart" button. Consumer electronics are a prime example, due to the wide range of features and functionality they offer, but nearly any product should be considered.

> **Customers, sales and support staff should be notified whenever there is any change to any product manuals.**

Clothing manuals should refer to the company's sizing system, as well as product specifications such as materials, measurements, and care and usage instructions. Up-to-date, well-written manuals and on-screen descriptions should also be readily available.

Retailers might question whether there is any need for providing online manuals if they offer telephone or online support, but not all customers like to engage in online customer service chats, or to speak to a customer service representative on the phone. Many times the customer is well informed and knows what is needed, but they just want the retailer to point them in the right direction. In these cases, easy-to-find online manuals are a valuable sales tool.

Inventory control is vital, as is warehouse and online interaction cohesiveness when providing the exact manual of an out of stock product. Sometimes the inventory system does not present the manual of a product that is not in stock because it ties all the relevant links to one entity. Flexible technology provides the option where retailers can still present the manual of a product that is out of stock and also make the customer aware when the particular product will become available again.

3.6.3   Content Distribution

Since product content is efficiently stored and managed, the distribution to the supporting manuals is critical. Retailers should ensure a management process that co-ordinates with the warehouse, logistics, and customer to provide a seamless presentation of the latest versions required.

> **Manuals are effective tools to enhance your retail brand and also cross-sell and up-sell other products relevant to the main item already sold to customers.**

Systems should be smart to trigger alerts that inform sales when a new product is introduced without online help or whether an existing product is discontinued. These alerts should be carried one step further to inform the customer who has already bought the product to inform that such changes have occurred. This strengthens your customer relationship and the shopper feels confident in their purchasing decisions.

The content management system chosen by an organization need have the flexibility to accommodate manuals. Ideally, these systems should facilitate the more advanced editing and updating of many versions of product manuals.

Providing self-service capabilities to interested customers is a powerful technique in building loyalty and closing sales.

eCommerce Best Practices

3.7 Customer Call Center Support & Service

- 3.7.1 Introduction
- 3.7.2 Investment, Not Expense
- 3.7.3 CRM and CSR Training
- 3.7.4 Click-to-Call / Click-to-Chat
- 3.7.5 Maintaining Customer Service
- 3.7.6. Call Center Optimization

3.7.1 Introduction

Customer Service is the act of addressing customer concerns and problems before, durring and after a sale. It is one of the most critical facets of an organization's operations, and – managed correctly – can greatly enhance sales.

Implementing a large-scale customer service operation properly is no small task. Most organizations rely on one or more connected applications to structure and streamline their Customer Relationship Management – or CRM – system. The complexity of this system is dependent on the size of your customer base, number of products for sale, and the required number of customer service representatives needed. As these factors increase is size and importance, the required CRM system becomes more expansive, intricate and intelligent.

> Consider your customers and catalog when determining the scale of your customer service operations.

3.7.2 Investment, not Expense

Many companies consider customer service just another overhead expense or sunk cost that will never be recovered, and do not view it as an investment in customer retention and brand management. Treating customer service as a necessary evil is the wrong way to approach the matter, as this attitude will inevitably result in a diminished return on investment. Not only is quality customer service the right thing to do, but it serves an organizations best interests – and bottom line -- to keep customers happy. Happy customers return to spend more money, and tell their friends about their happiness as well.

eCommerce Best Practices

By making a concerted effort in hiring qualified customer service representatives, the likelihood of retaining not only more customers but repeat customers will increase. Loyal customers are one major aspect of how investing time into selecting the best representatives will pay off in the end.

Zappos.com – for example — are considered a service company that just happens to sell shoes and accessories. They have made their customers their number one priority, to which that attitude stemmed from the company's core culture and values. Because their focus was not explicitly to increase their profit margin, they opted to use the money allotted for Marketing and put it into the funds for making free overnight shipping possible. In addition to that, Zappos' investment in customer service is shown by not limiting the CSRs to looking at their site. If a product is out of stock on the Zappos site, the CSR will direct customers to competitors' sites in good nature. Though an unorthodox approach, they are banking on them returning to Zappos the next time they need to buy shoe or accessories.

The Zappos model is a case of exemplary customer service, but understandably not every company can take the risk of making it a primary focus. However, by looking at Zappos' growth history in conjunction with their quality customer service and reliance on word of mouth, it is clear to see that it pays to offer great customer service.

3.7.3 CRM and CSR Training

Customer service representative are the voices representing a company, no matter how big or small it may be. The importance of these envoys being well-informed and well-trained is indisputable. They need to have access to pertinent information like; product technical and pricing details, SKU numbers, availability, shipping rates, and more. But a good customer relationship management system can enhance your experiences with customers as well as improve productivity with the CSRs.

An organized CRM system can manage all types of information necessary to facilitate quality service. Most CRM tools are customizable to fit your needs, whether you are looking to get customer insight, build customer profiles, design a similar CSR-facing UI that mimics the customer-facing UI, or generate reliable demand forecasts. Before you pick any one CRM tool, make sure you know what you need from the system first. You may not need every function available

eCommerce Best Practices

with the tool, because they may need a longer learning curve for employee adoption. Many times the best option is the most simple.

CSR training is a big factor in the effectiveness of your customer service, but sometimes even more important than training is hiring people with the right attitude. Any person has the ability to be trained, but not everyone adopts the right approach when interacting with customers. That being said, training is still very important for CSRs because their job is not only to assist the customer but to do it in a timely manner, and without proper training customers can lose patience very quickly.

Up-selling and cross-selling are two of the biggest contributors to increasing an order size or facilitating higher conversion rates. To do this effectively, CSRs need to have access to each user's profile in order for them to know what types of products can incite interest in their cross or up-sells. Not to say that CSRs should do hard sales pitches, but customers like to have more options than no options whether they realize it or not. Cross-selling and up-selling are ways to provide customers with relevant items based on the purchase they are considering making or currently ordering.

> **Training and attitude are key when selecting Customer Service Representatives.**

3.7.4 Click-to-Call / Click-to-Chat

Click-to-Call and Click-to-Chat technologies allow give customers the ability to make immediate contact with a CSR via phone, VoIP or Instant Messenger. An advantage of customers using either of these "alternate" methods of contacting a representative is that it is extremely convenient because they are already using a computer and are already online at the site. With just a click of a button, they can have live assistance from somebody on your team.

One other benefit of Click-to-Chat is that it enables a CSR to communicate with more than one customer at a time. Different chat boxes can be managed singularly yet simultaneously. The skill-set for a CSR on chat has to be slightly different because typing capabilities and familiarity with the chat technologies should be preferred.

A Click-to-Call option gives users an opportunity to talk by phone, in which the user would click on a link which notifies the CSR that he or she wants to communicate by phone and subsequently the rep would place a call to the given number in order to communicate with that user. The user also has the

eCommerce Best Practices

opportunity to call via PC, in which the user clicks on that link, has his or her PC scanned by the system for a microphone and speaker, and is thus connected to a CSR. Like Click-to-Chat, most these tools do not need to be downloaded or installed on the users end.

Being able to integrate this tool within your eCommerce environment and across channels is equally important as having the tool itself. You want to be able to draw up the user's profile if they have shopped at your site before or you want to be able to create a profile if it is the first time they are making contact. Making this contextual, real-time based information readily available is integral to what the CSR can cross-sell or up-sell. Being able to access the customer's shopping history can create greater opportunities for your reps.

> **Customer Service Representatives need quick access relevant user information and history in order to provide better service.**

3.7.5 Maintaining Customer Service

The impression you make with customer service, whether good or bad, stays with customers for a long time. Of course, it is your job to stay on the good spectrum of things because customers are known to hold grudges if they believe someone from your team committed a disservice. And only if you affect one person negatively, that person tells five friends and they tell five friends and soon enough you end up with large chunks of the population with a grudge against you. It is important to steer clear of poor interactions and take the necessary steps to make your customer happy. The customer may not always be right, but it is your job to make them think they are.

3.7.6 Call Center Optimization

Implementing Call Center Software, hiring CSR staff and managing user problems are all foundational characteristics of a fundamental Call Center Support Center. It is managements role to maintain and instill a layor of key metrics that will measure performance from a quantitative perspective. Some of the market leading KPI's include the following:

> **Wait Time** – Length of time a CSR is waiting for a call to be received. Popularly known as Down Time.
> **Talk Time** – Length of time that a CSR is physically engaged with a single call.

eCommerce Best Practices

Wrap Time – Length of time that a CSR spends after Talk Time to document the customer issue and classify it in the Call Center application.

The goal of CSR management should be to marginalize the summation of Talk Time and Wrap Time. This metric is invaluable to measuring and pin-pointing any CSR performance bottlenecks.

IV

technology

TECHNOLOGIES

The technology available today to power eCommerce sites is more efficient than ever before, and can be adapted for any organization's particular needs. For example, companies now have the capability to fully engage and interact with customers through such unique tools as Click2Chat or Voice over IP(VoIP). Also, newer architecture advancements allow organizations to realign their IT assets into self-operating departments specifically-designed to support profitable departments. Regardless of any technological tools to select from, it is the organization's ability to execute these new applications and methodologies that truly measures their effectiveness.

In this section, you will learn about some the following tools that are available when building or upgrading an eCommerce platform:

- The benefits of providing a positive, memorable user experience
- Useful tips for disparate application integration
- The importance of information architecture
- The relationship of product and service vendors
- Security concerns
- Infrastructure considerations
- The advantages of analytics and business intelligence
- Personalization
- Search capabilities
- How chat is affecting online communication
- Customer influence with blogs
- How syndication is reshaping information exchange
- The benefits of portals

eCommerce Best Practices

4.1 User Interface / User Experience

- 4.1.1 Introduction
- 4.1.2 Considerations
- 4.1.3 Best practices

4.1.1 Introduction

First impressions create lasting expectations, and online shopping is no different. The look and feel of your site is crucial to success because that is what consumers intrinsically respond to the most. From the consumers' perspective, strong impressions are quickly formed based on your site's user interface (UI) and how deep your user experience (UX) goes.

> **Perception is reality. You need to build a system which is not only really good on the inside but which also looks really good as well.**

An eCommerce system is only as good as how a visitor or user of the system perceives it. As in many other fields, perception is reality. The way a user perceives the system translates to real value and equity. Even though it is important for a site to be architecturally sound with a stable and competent infrastructure, when potential customers visit your site, they know nothing about the system internals. What they are concerned with is that the site is intuitive and looks good; and what you should be concerned about is that the site is built in a manner that it elicits positive user response.

This chapter will present some UI and UX related matters and will suggest some recommended best practices that can be applied across industries.

4.1.2 Considerations

People have their own sense of style, their own particular value system, and various cultural influences that determine behavior patterns and reactions to specific kinds of stimuli or situations.
If you have a specific target demographic, it is important that these aspects are understood, and are considered in designing your eCommerce site's user interface.

> **Consider your users needs and preconceptions when developing a User Interface.**

The web interface is not only a reflection of the desired user perspective, but also reflects the various business needs. It is important to present familiar elements to the user, may it be standard navigation, workflow processes, personalization elements or promotional items. Unfamiliar and non-intuitive elements on a web page will create an un-necessary learning curve for the customer, frustrate their shopping experience and result in fewer sales.

Surveys consistently indicate that usability is by far the greatest factor in determining the success or failure of an eCommerce system. An enriched, elevated, and personalized user experiences will deliver repeated visits from the same user and would allow the site to gain a positive reputation through word-of-mouth.

4.1.3   Best Practices

The following lists some key recommendations for best practices in User Interface and User Experience design. While not exhaustive, it provides a general guideline that will be useful in making specific and detailed guideline for any eCommerce system.

Presentation & Content

- Create a system that is as intuitive as possible for end users to interact with and navigate.
- Enhanced product displays with use of Rich Media create a real time visual feel and increases customer and product interactions. Furnishing products could have larger displays while apparel could be presented showing the backs and fronts.
- Page elements that are frequently used like shopping cart, back buttons, help, FAQs, and Search should be in a consistent location throughout the site to ensure smoothness in customers' navigation flow.

> **Consider presentation, content, system attributes, business processes, and customer relationship in making the system user friendly and user oriented.**

eCommerce Best Practices

- Navigation flow should be based on frequent customers' shopping patterns and should mimic the different types of activities a customer would perform in a store. At a very high level these are making purchases, returns, product information/guidance (FAQs) etc. At a more detailed level it would be purchases /browsing to different sections e.g. DVD section/book section/children section etc.
- Create and/or use various patterns that customers can easily remember.
- Stick with the best-of-breed sites; reuse elements which have resulted in better conversion rates in the past.
- Excellent and objective content – such as buying guides – are simple and cost-effective techniques that gain customers' trust, build customer relationships, encourage repeat visits and increase revenue.
- Thoughtful product descriptions, innovative imagery and clearly presented product pricing close the gap where a customer would want to touch, inspect and try the product.
- Provide convenience to customers. Put user entered data in sessions so that he they do not have to type in the same information repeatedly in the same session. Make the customer enter the least amount of data possible and try to capture more data implicitly through user actions.
- Provide seamless cross-channel customer experiences. These may include cross branding, inventory management and customer focused pick-up and return policies promote trust, purchasing comfort, goodwill and an excellent customer service experience.
- Group associated products, and make cross-sell suggestions to customers. People like to see products in scale or grade. Also, display comparisons between different products as shoppers like to weigh out their options before making a purchasing decision.
- Display whether a product is in-stock, out-of-stock or almost out-of-stock.
- Clearly show the price breakdown.
- Allow your customers to change their order at any reasonable point in time.
- Be liberal in showing catalogs. Generally speaking, the customer should have visibility to all your product catalogs without needing to input too much information in the search box.

Important Design Goals

- Build and portray a strong sense of system security and build confidence in customers by assuring them that their private data will not be

eCommerce Best Practices

compromised. Use third-party security vendors like Scan Alert or VeriSign to increase credibility.
- The integration between various functional domains and system domains should look as seamless as possible to the customer.
- Make the system expressive. Highlight required fields and tell customers up-front if there are any issues in their order. Display links to security and privacy policies and avoid general error messages where possible.
- Make the system flexible in user registration and ordering process so that customers are not unnecessarily delayed when trying to achieve their final objective. For example, present customers fewer pieces of information to fill in while registering in the checkout flow. This will allow a customer to quickly make their purchase, and let them fill in the other fields the next time they login.
- "Search-and Serve" should complement each other. Shoppers should be able to get the right products in their search results. Help the shopper by automatically correcting spelling mistakes and by creating synonyms for popular search terms (e.g. a search for Brittney automatically searches for Britney).

Keep the Customer in Mind

- Always get a confirmation from the customer, inform them of order status, and communicate (via email) with the customer about his or her purchase. This gives customers assurance about the selection.
- Ask a limited number of questions from the customer while registering. You can implicitly capture more information based on the customer's actions and shopping behavior.
- Ask only critical questions up-front, saving additional questions for later. Shipping and billing address, for example, need only be asked after the customer wants to make a purchase.
- Use the same information as much as possible. If the shipping address and billing address are the same, do not ask the customer to enter the same address twice. Ask the user if the two are the same and if the user confirms that, automatically populate data from one field to another.

> **Users will reward investment in a user-oriented system with increased purchases.**

- Explain benefits of registration like personalized web experience, access to the customer's order history, more expedited future checkouts, etc.
- As much as possible, present multiple choice-type questions for getting the customer's responses.

- Allow the customer to easily un-register and re-register at any reasonable point in time.
- Provide ample opportunities for the customer to sign-in, but also allow him or her to work anonymously.

The fundamental goal of your eCommerce system is arguably a positive user experience. It is not just a side note or a formality — it is one of the most important factors that determine your site's longevity. By providing a good user experience you will not only get a short-term return in investment via higher sales, but also a continuous long-term return via repeat visits and purchases and personal recommendations to other potential users by satisfied customers.

4.2 Integration

4.2.1 Introduction
4.2.2 Data Integration
4.2.3 Web Services, SOAP, WSDL, and UDDI
4.2.4 Services Oriented Architecture
4.2.5 What is SOA?
4.2.6 Implementing SOA Using Web Service
4.2.7 Enterprise Service Bus
4.2.8 ESB Implement Requirements
4.2.9 ESB and Web Services

4.2.1 Introduction

Application integration and interoperability are essential for establishing functional business processes. Because this topic is a whole subject in itself, this section will only provide a high level detail of some popular integration methodologies.

Over the years, businesses and institutions have used/established various software applications in order to address and solve problems for each of their business needs. Different software applications have emerged to address the various requirements of different business domains e.g. Human Resources, Customer Relationship Management, Manufacturing Resource Planning, Warehousing, Supply Chain Management, Financials, etc. However, these software applications have mostly been developed and used in isolation – as "silos" – addressing only their own specific business domain. In order to realize their full potential, these software applications must be interconnected and seamlessly integrated with each other, by establishing interoperability, both syntactically and semantically. The following sections describe some of the most common ways and techniques used for addressing integrated and interoperability issues.

> **Software applications should be interconnected and integrated in order to establish interoperability.**

4.2.2 Data Integration

A decade ago, data integration was the most common technique for integrated software applications to establish interoperability and application-to-application communication. Since the 1960s, data integration has been established through data warehousing, in which software application data is extracted, transformed, and loaded (ETL) from multiple software applications' source databases to a big and common data warehouse (database), thereby making it possible to exchange and reuse data between them. Over the years, data integration moved to mediated (virtual) schemas. In the market, various vendors provide development tools to establish data warehouses. Data warehouses are established through scripting, data transfer, data replication, data mirroring, disk mirroring, etc. Due to such complexity, rapid or quick changes to business needs are hard to realize. However, data warehouses are excellent repositories for analytics and business intelligence tools.

4.2.3 Web Services, SOAP, WSDL, and UDDI

A popular method of integration is to exchange data at an application level as opposed to transforming large amounts of data at the database. Known as Enterprise Application Integration, or EAI, this level of integration is typically focused on business process integration and managing data traffic between disparate applications via messaging interfaces. However, these interfaces are not designed to handle large volumes of data and are typically suited to enable business processes in real time.

With the introduction of SOA (Service-Oriented Architecture), technical breakthroughs and advancements in XML have created an environment that has spearheaded the integration process through Web Services and ESB (Enterprise Service Bus).

Web Services
Web Services are new, exposed application programming interfaces (API) that can be accessed over a network and executed on a remote system hosting the requested services. For example, from an eCommerce perspective, order management through disparate applications becomes synchronized through Web Services, much like EAI (Enterprise Application Integration) was used in the past.

> SOAP is supportable in many existing environments and can be transported over a variety of protocols.

eCommerce Best Practices

Web Services grew from adapting to how people and systems interact, enabling new business models, extensions of opportunity, new transparency and improved collaboration between employees and employers, and in some cases reductions in infrastructure costs. The key to these successes was a universal server-to-client model that is consistent with a highly distributed environment, based on simple open standards. Web Services provides ways for disparate systems to communicate by integrating one business directly with another so that the process does not have to wait for people to provide the glue.

Although Representational State Transfer (REST) based Web Services are possible, due to common use and its prevalent nature, SOAP based Web Services are described in this book and recommended as the best practice.

SOAP: Simple Object Access Protocol
The most straight forward way of describing SOAP is to say that it provides clear communication two business applications that "speak" different languages. SOAP is an XML messaging protocol that is independent of any specific transport protocol. By defining a framework for messages to control the behavior of SOAP-enabled middleware, and a message body it is able to allow disparate applications to operate smoothly. SOAP is supportable in the vast majority of existing and new technical environments and can be transported over a vast variety of protocols.

SOAP also makes no reference to characteristics of interactions such as security and transaction. However, as SOAP messages provide a highly future-flexible model, these aspects are being added gradually to the Web Services specifications as extensibility elements.

WSDL: Web Services Description Language
WSDL is an XML-based interface definition language that separates function from implementation, and enables design by contract as recommended by SOA. WSDL descriptions contain a port type (the functional and data description of the operations that are available in a web service), a binding (providing instructions for interacting with the Web Service through specific protocols, such as SOAP/HTTP), and a port (providing a specific address through which a web service can be invoked using a specific protocol binding).

The value of WSDL is that it enables development tooling and middleware for any platform and language to understand service operations and invocation mechanisms. For example, given the WSDL interface to a service that is implemented in Java, and offering invocation through HTTP, a developer working in the Microsoft .Net platform can import the WSDL and easily generate an application code to invoke the service. Along with SOAP headers, the WSDL specification is extensible and provides for additional aspects of service interactions to be specified, such as security and transaction.

UDDI: Universal Description Discovery Integration
Web Services established using SOAP-formatted XML envelopes that have their interfaces described by WSDL, can sometimes be found using UDDI. UDDI servers act as a directory of available services and service providers. SOAP can be used to query UDDI to find the locations of WSDL definitions of services, or the search can be performed through a user interface at design or development time.

4.2.4 Services-Oriented Architecture (SOA)

SOA is a method of organizing your organization's IT assets by linking business and technical resources. Applications collaborate by invoking their independent application services. These services are composed into larger sequences to implement business processes as software application solutions. If correctly implemented, the fundamental benefit of SOA is that it allows the business to more quickly and cost-effectively adapt to changing market conditions.

The development of Service Oriented Architecture has created a more uniform and flexible application environment across multiple business processes. The act of separating different business functions into distinct service units is the essence of SOA. What makes it so useful is that it because these services are available on a shared network, they can be recycled or reorganized to make a variety of business applications built from software services.

> **There should be a consistent mechanism through which services communicate.**

One of the primary benefits of SOA is that it takes relatively large groups of functionality and strings them together to create ad-hoc applications. This is believed to be the future of business processes because the services can be optimized fairly rapidly and at lower costs.

In order for a Service Oriented Architecture to run properly, you need to make sure your system can function in that environment. Before integrating the

 **eCommerce Best Practices**

applications across different platforms, be sure that you have interoperability from system to system and different programming languages. You will also want to collect system information into a data warehouse, to give new functionalities access to standardized information.

By adopting a SOA approach and implementing it using supporting technologies, you can build flexible systems that implement changing business processes quickly and make extensive use of reusable components.

If a company already has retail, warehouse, supply chain, and billing systems, it should build the new process by reusing the functionality that is provided by those systems rather than having to write new applications or new interfaces to the existing systems. If the company has already adopted a SOA approach, it will have defined the interfaces to its existing systems in terms of the functions or services that they offer to support the building of business processes. The defined interface makes building the new web front end to the system very simple.

4.2.5 Enterprise Service Bus

Implementing a SOA requires applications and infrastructure that can support the SOA integration guidelines. Applications can be enabled by creating service interfaces to existing or new functions that are hosted by the applications. The service interfaces should be accessed using an infrastructure that can route and transport service requests to the correct service provider. As organizations expose more and more functions as services, it is vitally important that this infrastructure support the management of SOA on an enterprise scale.

What is ESB?
The Enterprise Service Bus (ESB) is another middleware infrastructure component that supports the implementation of SOA within an enterprise. An ESB provides an infrastructure that removes any direct connection between service consumers and providers thereby providing a scalable and integrated architecture. Consumers connect to the Bus and not the provider that actually implements the service. This type of connection further decouples the consumer from the provider. A Bus also implements further value added capabilities. For example, security and delivery assurance can be implemented centrally within the Bus instead of having this buried within the applications.

4.3 Service Oriented Architecture (SOA)

- 4.3.1 Introduction
- 4.3.2 SOA Key End Points
- 4.3.3 Suitability
- 4.3.4 Guidelines
- 4.3.5 Benefits of SOA
- 4.3.6 SOA and Web Services
- 4.3.7 eCommerce and SOA

4.3.1 Introduction

Service Oriented Architecture (SOA) is one of the new IT methodologies that have caught main stream attention in the last few years. And because of this, in the broader sense, there is some confusion on what it truly means. Some literature has taken the literal meaning of service orientation and treated as more of a mind-set. Others equate very specific technologies to Service Oriented Architecture. And, finally, there is a group which defines it in terms of a strict pattern where the architecture contains the distinct roles of services, services directory and service consumers. This chapter will define meanings of Service Oriented Architecture, help you evaluate if it fits with your organization, and suggest some industry best practices.

It is important to understand what each person involved in the SOA initiative understands about the concepts involved and get everyone on the same page and, if required, change the terminology of the initiative to reflect the actual task. Terminology change might be required to prevent confusion for new team members joining the company or for job postings in order to target the right people for the job.

The most popularly accepted definition of Service Oriented Architecture is an architecture where functionalities are made available as discoverable and loosely coupled services in a platform independent manner. In SOA service, the consumer is not dependent on and does not need to know the underlying technology used for implementing the service. This allows functionalities to be exposed throughout the organization and allows functionalities to be used very quickly between different departments of a company or with partners. This can be implemented using different technologies including Web Services, REST, RPC, DCOM or CORBA.

It is important to note that sometimes the term SOA and Web Services is used interchangeably. Web Services do provide all the necessary components which can be used to implement the SOA strategy in your company. However web services are not the only method to implement SOA in your organization. Other technologies such as REST can fulfill some of the requirements of SOA.

4.3.2   SOA Key End Points

The three main architectural end points in SOA are defined below:

Service Provider: This is the implementation that makes available the underlying functionality to any consumer who requests it. This functionality could be as discreet as performing an inventory lookup or involve a more complex business process such as submitting an order. The interface exposing the underlying functionality could use a different number of technologies, such as Web Services, REST, CORBA etc.

Service Consumer: The Service Consumer as the name indicates is the application which actually uses the information requested from the service provider.

Service Directory: The Service Directory (which is also known as the Enterprise Service Bus), is the mediator that facilitates the service consumer and provider. The service provider registers itself with the directory while the consumer submits requests for services with the directory. The Service Directory informs the consumer of the right service, which the consumer then communicates with to request the necessary functionality/information.

The Service Directory may also act as a bridge to legacy applications, in which case the consumers would receive all the information from the directory itself. For the Service Directory to act as a an efficient bridge, it must provide functionality to communicate using open standards (JMS, Web Services) as well as provide adapters to enterprise data sources (such as SAP, PeopleSoft, Oracle SQL Server). This allows for the consumer to have just one interface to the Service Directory as opposed to implementing different interfaces to each service.

It is also important to note that if not all the elements of web services is used (such as the Service Directory or Enterprise Services Hub (UDDI)), then it fails to meet the criterion of SOA and is just a standalone service which is publically available for usage.

4.3.3 Suitability

Not every company needs an SOA solution. There are a number of factors which determine if a Service Oriented Architecture is the right fit for your organization.

Adopting a Service Oriented Architecture only makes sense when there are a disparate set of applications which run in your organization which cannot communicate with each other, or if there are plans for applications which may make available information for consumption. Of course it goes without saying that the integration of the disparate applications should fulfill a concrete business goal and produce value for the company itself.

If there are just two or even three applications which need integration without any new foreseeable applications coming into your organizations, then SOA might be overkill for your company. Furthermore, it is only seen as being a wise investment for companies who are middle to large size.

4.3.4 Guidelines

The implementation of a Service Oriented Architecture is not a small endeavor and can be a very expensive affair. It involves implementation/configuration of Service Directory (or ESB), and making available the desired information from various systems. Exposing each of the different information silos might require a team of different skill sets to implement. If the team involved in the implementation of exposing the services is not the same team as the original implementers, then additional time would be required for requirements and analysis. Needless to say, each service that is made available is a project on its own. After one or two of the services are exposed and are able to register themselves with the Service Directory, the work to have the Service Consumer successfully use both the directory and the service still need to be done to finally prove the validity of the implementation.

Consider the alternative technologies and strategies that might be used instead of SOA to achieve your business goals. This is not to discourage you from using SOA, but would put into perspective the benefits of implementing an SOA solution in your organization. It might turn out that SOA is overkill or alternatively that it does reduce cost in the long term. Performing such an architectural evaluation will bring to light the timelines required and make sure

that you are heading down the right path. It is better to explore the alternatives in the beginning than lose an investment due to adaptation of the latest buzz-word.

Given the time and expense of an SOA implementation it would be best to start with a proof of concept. Some of the benefits of doing a small proof of concept are:

- Shows concrete business value to your organization,
- Better estimate the cost involved in future implementation,
- Generate business buy-in for larger phases for the project,
- Gives an opportunity to consider and optionally evaluate alternative SOA related technologies for implementation,
- Helps identify pitfalls in SOA implementation,
- Brings all the developers up to speed on the technology

4.3.5   Benefits of SOA

Reusability: In a typical organization, different departments would have developed different functionalities to meet their needs. If the same business function needs to be used by another department, then this other department would need to either implement a direct link with the original application or re-implement the functionality. By using SOA principles, exposed applications can be reused across multiple departments in the organization. Similarly each department would only need to create one interface (with the Service Directory) to be able to make calls to any number of exposed functionalities across the organization.

Increasing Business Agility: By applying the SOA principles to your organization, the main business functions would be made available. If and when there is a need to expand the business into a new area or to integrate with another system, the initiative would be faster than without an SOA infrastructure in place. Most likely the existing services would serve to provide the necessary functionalities for the new initiative or the necessary ones would be exposed for future usage. And the new application, once exposed via services and service directory, would provide the necessary functions/information to the rest of the applications.

Governance: The Service Directory or the ESB can make for a central location to enforce policies and control the functionalities that are accessed and information which is exchanged between different systems. This is also commonly known as SOA Governance. The Service Directory can be used to enforce security compliance as required by different standards, mask data which shouldn't be shared with certain departments, refer the right service amongst a set of services which provide the same functionality and handle permission to access the different services.

Flexibility: Because different functionalities are made available across departments, they can be aggregated together in a new service to provide new functionalities. These results in faster time to market to quickly meet any new demands placed on the organization.

Cost efficiency: Since functionalities are exposed in the organization, the extra work and cost of re-implementing some of the logic can be avoided.

4.3.6   SOA and Web Services

As mentioned previously, SOA and Web Services are often taken to mean the same thing. However there are some exceptions to this case that require clarification. It would be accurate to say that SOA is the architectural approach to company-wide integration and Web Services provides one concrete set of technologies to achieve this goal. This section will expand on how Web Services actually map to the SOA approach to your company.

XML: The markup language that contains the information which is sent between the provider, directory and the consumer.

Web Services Descriptor Language (WSDL): This is the standard XML format which is used to describe the interfaces available for invocation. It also describes the protocol binding and message formats required to interact with the web service. These map to the common interface which is published by the services for the consumers.

SOAP: The protocol for exchanging the messages between provider and consumer.

Universal Description, Discover and Integration (UDDI): This component maps to the Services Directory in SOA. UDDI is used to find the right web service for the consumer and notifies them of the interface available via the correct WSDL.

eCommerce Best Practices

4.3.7 eCommerce and SOA

SOA is an architectural approach which may or may be needed for eCommerce. However, it can certainly leverage your eCommerce site or be leveraged by it. When implementing your eCommerce system your retail related functionality should be exposed (if they exist) for usage to allow for one central place of control. Otherwise change in your existing system would have to be repeated in the eCommerce system as well. Some of the functions that might be made available in the eCommerce site are given below.

- Order Submission
- Order Status
- Catalog Lookup
- Search
- Profile Creation
- Profile Updates
- User Activity
- Inventory Lookup
- Credit Card Authentication

Some of the functionality which an eCommerce site might expose for other applications in your system can include:

- Carts in pre-submitted state
- Abandoned Carts (for Marketing)
- User Activity (for Marketing)
- Wish/Gift/Registry List queries (for Marketing)
- Order Lookup (for CSR)
- Profile Functionality (for CSR)

An important note is that while following an SOA approach may fulfill departmental needs regarding eCommerce, if not done correctly, unacceptable overhead might be introduced into the system which may hinder the performance of the eCommerce system. For example a marketing query into user activity might produce some undue overhead on the eCommerce database and effect user experience. The right process would be run database replication on the main eCommerce database and then expose the different types of reports via services. In other cases a direct lookup by a CSR might not create this issue (given there isn't a large number of queries).

eCommerce Best Practices

4.4 Application Architecture

4.4.1 Introduction
4.4.2 One-Tier Architecture
4.4.3 Two-Tier Architecture
4.4.4 Three-Tier Architecture
4.4.5 N-Tier Architecture
4.4.6 MVC

4.4.1 Introduction

The success of the Internet is due to its software architecture, which has been designed to meet the needs of distributed applications in heterogeneous environments. The Web has been iteratively developed through a series of modifications of the standards that define its architecture. The modern Web architecture emphasizes independent deployment of components, scalability of component interactions, generality of interfaces and intermediary components to reduce interaction latency, enforce security, and encapsulate legacy systems. In this chapter you will learn about the different kinds of software architectures that emerged with time to fit the current day web requirements.

4.4.2 One-Tier Architecture

The simplest and least extensible of all system designs, the one-tier architecture is a standalone application program that runs all aspects of the system. Both the system memory timings and user interface requirements all occur at this location. Again, a highly inefficient model and one that is ancient to say the least.

> **Two-Tier architecture takes less time to build versus other architectural models but is typically geared towards simpler applications with few users.**

4.4.3 Two-Tier Architecture

The two-tier design of applications consists of the three components of an application: presentation, processing logic and data—distributed into two layers among the client and server. In this design, presentation is handled exclusively by the client, whereas the business logic required for processing the data is split between the client and the server. All application codes reside on the client side and the data on the database server.

One of the major advantages of two-tier architecture is the speed of application development. The time required to develop a two-tier system is considerably less than the time required to develop a system with n-tier architecture.

Limitations of Two-Tier Architecture

The two-tier architecture is suited for relatively simple, non-critical applications with very few (< 100) users mainly because the scalability becomes an issue for such systems. Also, any change in the business rules or change in the database management system requires a costly rewrite of the application. Two-tier applications become an administrative nightmare whenever there is an upgrade. All the updates have to be delivered, installed and tested on every client. This might lead to lack of uniformity of the client application if there is no consistent methodology developed specifically for system upgrade.

4.4.4 Three-Tier Architecture

The three-tier architecture emerged in the 1990s to overcome some of the limitations faced by the two-tier architecture. This architecture added a middle tier as the new third tier in between the client side and data management components. The middle tier serves as a business layer where all the business rules are executed, thus, differentiating the presentation layer from the business logic. The promising features of a three-tier architecture are increased performance, flexibility, maintainability, reusability and scalability—all provided while hiding the complexity from the user.

> The middle-tier differentiates the presentation layer from the business logic, making a place for executing business rules.

A majority of the current Internet based applications are built upon a three-tier architecture model comprising of the user interface layer (client tier), functional application logic layer (business logic tier) and the data storage and access layer (database tier).

The user interface is the means by which the end users of the application interact with it; it provides means of input and output. The most common user interface medium is a web browser which accepts input and provides output by generating a series of web pages. The user interface can also be any other client based Graphical User Interface (GUI). The user interface has very little or no application logic built into it.

eCommerce Best Practices

The business logic layer comprises of the business rules that define the application and handles the workflow between the other two tiers of the three-tier architecture model. This layer improves performance, maintainability and scalability by handling all the process logic. Any changes must be written and deployed only once in the middle-tier to be available to the whole system.

The database tier consists of a database management system where the data is stored and managed. This layer ensures consistency of the data across all users.

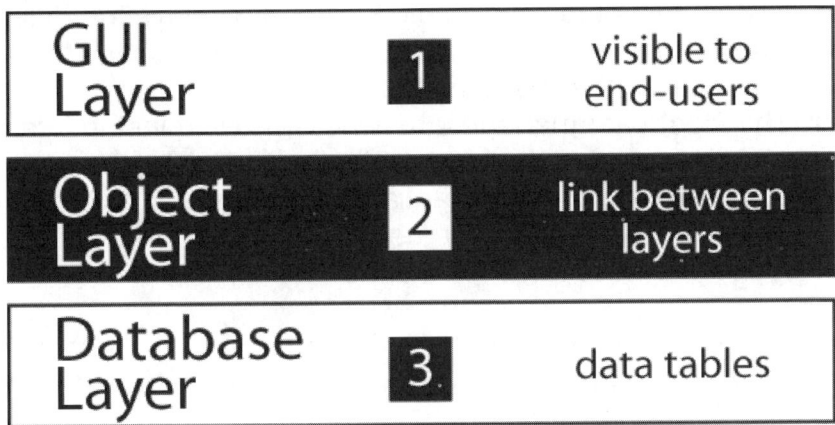

4.4.5 N-Tier Architecture

N-Tier architecture is typically divided into four components or layers: Presentation, Business Logic, Data Access and Database. Each layer operates independently of each other and requires interface by various users and applications. For example, the client, or user, generally is free from interacting with any layer other than the presentation layer. This would consist mainly of operating system interfaces like Windows or DOS. The business logic and data access layers combine to operate on an application server, while the database layer rests on a database server typically of a relational format.

4.4.6 MVC Architecture

If a component or piece of functionality interacts directly with both the user interface and the database, then it is very sensitive to changes at every level of the system. In order to reduce the need to constantly update many

eCommerce Best Practices

components, three classes of components are designated to distinguish the data presentation from the data maintenance layer, and have a third component that coordinates the first two. These three components are called the Model, the View and the Controller and they form the basis of the MVC Pattern.

The responsibilities of the basic MVC components in brief:

Model: The Model is responsible for keeping the data or the state of the application. It also manages the storage and the retrieval of the data source. It also notifies all the views that are viewing its data when the data changes.

View: The view contains the presentation logic. It displays the data contained in the Model to the users. It also allows the users to interact with the system and notifies the Controller of the user's actions.

Controller: The Controller manages the whole show. It instantiates the Model and the View and associates the Model with the View. Depending on the application requirements, it may instantiate multiple views and associates them with the same Model. It listens for the user's actions and manipulates the Model as dictated by the business rules.

From an eCommerce perspective, MVC delivers the best content and style management. Meaning that it is the most flexible and adapable to your business goals and demanding results.

Advantages of MVC Architecture

Separating the application behavior (Controller) from the data presentation (View) allows the controller to create an appropriate View at runtime based upon the Model.

Separating the data representation (Model) from the data presentation (View) allows multiple Views for the same data. Changes can occur in both the Model and View components independently of each other as long as their interfaces remain the same. This increases the maintainability and extensibility of the system.

> **Because the Model is separated from the View, you are able to have multiple views of the same data, which increases the extensibility of the system.**

Separating the application behavior (Controller) from a representation (Model) allows the user's requests to be mapped from the Controller to specific application-level function in the Model.

eCommerce Best Practices

Although MVC involves communication between Model, View and Controller components, it is not a behavioral pattern because it does not specify how the three components should communicate. The MVC pattern only specifies that the structure of a system of components be such that each component take up one of the three roles — Model, View or Controller, and provide functionality only for its own role.

In the J2EE world, MVC is thought of more as architecture rather than a design pattern. The MVC architecture needs to be complemented with some design patterns at various levels, the presentation and presentation to business, and the business layer and data access layer. Design patterns induce abstraction, division of labor and reusability in software systems. Consistent use of design patterns results in a scalable and maintainable system.

4.5 eCommerce Platforms

- 4.5.1 Introduction
- 4.5.2 Perpetual License Vendors
- 4.5.3 Considerations before Selecting Service Vendors
- 4.5.4 Selecting the Right eCommerce System
- 4.5.5 When Product and Service Vendors Collide
- 4.5.6 PayPal, GoogleCheckout and Small Business

4.5.1 Introduction

Building an eCommerce site is truly a layered process, not just in the infrastructure and architecture of the site, but by all the different vendors that can be involved in developing and running the site. Product and service providers are available to help you create or maintain your eCommerce site so that you can focus more on the business side of operations instead of the technical side.

A lot of companies try to build their site internally using their own resources, but often they find that the extent to which they would like their eCommerce business to grow is not something they can create on their own. In this day and age, where eCommerce sites deliver 100% of some companies revenue, it is worth taking the time to research and invest in the different product vendors available to you.

4.5.2 Perpetual License Vendors

> **Selecting the right eCommerce system can give you more time to concentrate on the business side of operations.**

Product vendors can technically be any person or organization that offers something up for sale. This certainly does not narrow things down because the number of products out there is endless, but focusing on the types of eCommerce product vendors, specifically, will provide a relative shortlist of options.

Product vendors in eCommerce can offer eCommerce platforms, hardware, software, applications, integrated merchandising, analysis and CRM systems, etc. There are a myriad of choices for what system is best for your company and which applications will suit your needs. Keep in mind that many of the products that the vendors provide can be customized in order to meet your standards.

4.5.3 Considerations before Selecting Service Vendors

It is vital for project success to choose the right service vendor for the type of service you want. It is important to make sure that you clearly outline your requirements, penalties, and goals for the tasks that need to be completed. These are known as the Service Level Requirements, SLR, and the Service Level Agreement, SLA.

> **Specify your exact requirements in the Service Level Requirements document and the penalties involved in a breach.**

SLR

You need to draft a Service Level Requirements document that specifically states your expectations and objectives for the project that the service vendor has to adhere to. Not only does this act as a frame of reference for your needs once the project begins, it also protects your best interests when the service vendor is in place. It is important to be specific and thorough, leaving no stone unturned because complacency will lead to confusion, and ultimately, disappointment.

SLA

Once the Service Level Requirements document is submitted to the service vendor, it is then up to the service vendor to determine whether the requirements are in their reach. If it is deemed feasible and acceptable, a Service Level Agreement, SLA, is then made. This is basically a contract between you and the service vendor about the stating how involved the service vendor will be and what they are responsible for. Essentially, it renders the accountability to the service provider in completing the project.

Penalties

Should the service vendor stray from the requirements in the SLA, you should clearly state the penalties for doing so. This keeps the service vendor on their toes and holds them liable for their mistakes or negligence. Penalties can range from warnings to monetary fines, depending on the severity and/or frequency of the infraction(s).

eCommerce Best Practices

4.5.4 Selecting the Right eCommerce System

One of the biggest and most important decisions you may need to make is what eCommerce system will meet your online sales needs. An eCommerce system or platform is basically what makes the site function correctly—in terms of searchandising, content management, sales, payment processing, etc. It is the engine, so to speak, of your eCommerce site and you need to make sure that the engine is compatible.

Questions to Ask Yourself

- How much do you want to invest in your eCommerce site? This not only includes your software budget, but also your time in development and support, internal or external staffing.
- What kind of system will meet your customers' needs?
- How the system will reflect your offline capabilities – in terms of inventory management, customer service, pricing, etc.

Longevity

The online market is not going to go anywhere in the near future. It is here for the long haul, so you cannot afford to neglect your eCommerce site's potential. Depending on your company's size, number of SKUs, or customer base you need to evaluate the system that will ensure your online longevity.

Sample eCommerce Platform Vendors

- ATG Commerce
- Escalate Retail / Blue Martini
- IBM WebSphere Commerce
- Sterling Commerce / Comergent

> Keep in mind that you can usually select a system with different levels of capabilities, based on the size of your operation, as well as customize it to fit your needs.

Most of the eCommerce systems can be customized to fit your needs—whether you need content management, B2B, email or web marketing, merchandising, A/B split testing solutions, or a particular combination of solutions. Different systems are more user friendly than others, while other systems might be slightly more complex but can handle different operations.

4.5.5 When Product and Service Vendors Collide

Many times the product vendor is actually a service provider as well, especially if the product fits a certain niche need. Who better to provide the integration and support of the product than the vendor itself—they certainly know how product operates. For example, email marketing firms can offer their software in addition to its management and analytics.

Some companies might not have the resources to provide the service along with the products, but if they do it comes as a great advantage to you. One, you know you have knowledgeable service providers on hand to assist you, and two, you do not have to look far for that service.

Software as a Service (SaaS)

A new model of application delivery that is becoming more common is known as software as a service, or SaaS. Software used to be thought more along the lines of a product because companies would need to actually buy and own the software licenses in use as well as pay the maintenance fees in conjunction with the software, but SaaS has changed the way eCommerce systems operate. Now, software vendors can develop web based versions of the software to be delivered and operated over the Internet. After the software is implemented to your site, you would have not pay to own the software, but instead, they pay to use it (typically monthly fees based on applications usage). Depending on your business needs, a hosted solution may be a cost-effective way to get the most out of your eCommerce site while investing a fraction of what it would cost to own the software and hardware itself.

> SaaS is a great alternative to getting the solutions you need at a fraction of the cost. Hosted eCommerce is a growing trend.

There are benefits and drawbacks to using SaaS (also often called OnDemand or Managed Service Provider). Usually the up-front costs are lower and the implementation timeframe is faster with SaaS. You will not need to hire system administrators or developers for the site. However, the amount of customization and integration that can be performed with SaaS is often limited. Check with the vendor regarding what intellectual property you will own at the end of the service agreement (custom code, customer data, etc). Also consider the accounting implications of a capital expenditure for perpetual software licenses against the recurring expense of a subscription service.

eCommerce Best Practices

Popular MSP/COD/SaaS eCommerce venders include:

- ATG
- Demandware
- Fry
- GSI
- MarketLive
- Venda
- Yahoo! Stores

4.5.6 Pay Pal, Google Checkout and Small Businesses

Today, small businesses are experiencing greater returns because of a variety of service vendors that are bolstering their sales. The eCommerce environment has opened doors for many product and service vendors alike. Online shopping has reshaped the market in important ways, one being that it provides an alternative, global avenue for "mom and pop" stores to sell their products.

Pay Pal and Google Checkout are two types service vendors that enable everyone from individual sellers to small-run companies to make their products available on the market. Anyone visiting the service vendor site can search for the product based on keywords, product category, price, brand, etc. Consequently, the open environment of service vendors like eBay has carved out a customer base searching for obscure, novelty, over-stocked, and out of date products. This phenomenon is known as the "long-tail," which was spearheaded by the likes of Google and other eCommerce sites that give companies greater customer reach.

> **EBay and Google Checkout are good examples of online service vendors providing sellers and smaller companies a way to reach a larger customer base.**

Service vendors are here to assist you in reaching your goals, and in eCommerce there are a lot of vendors that can help you reach them. Be clear about the type of service and needs you want met because the more specific you are the more likely the vendor will be able to accomplish it. There are risks involved when deciding to enter in a relationship with a service vendor, but you can reduce those risks by evaluating your requirements and penalties of breach.

Small businesses are finding their spots in the market through eCommerce sites like StarbucksStore.com use of Pay Pal and Google Checkout. These service vendors have been able to expose more sellers and companies to a global and richer (literally and figuratively) audience.

4.6 Implementation Service Vendors

4.6.1 Introduction
4.6.2 Considerations Before Selecting Service Vendors
4.6.3 Advantages and Disadvantages to Service Vendors
4.6.4 Offshore IT Outsourcing
4.6.5 Managed Service

4.6.1 Introduction

One of the benefits of creating or enhancing your eCommerce site today is that you do not have to do all the work. Service vendors specifically related to eCommerce are here to take the burden off of having to learn about what will optimize your site's potential and having to search for qualified new employees.

A service can be defined as any discrete function that can be offered to an external consumer. The function can be an individual business function or a collection of functions that together form a process.

> There are many vendors for many services out there, especially relating to eCommerce. Take your time and evaluate several options.

There are many service providers for many types of services. If there is a task to be completed, there is a service provider available to complete it. In eCommerce in particular, there are systems integrators, content management services, email marketing services, web analytics services, fulfillment services, and many other specialized providers.

There are many different factors involved in selecting service providers. You can opt to get everything done in-house, which my require hiring specialists, or perhaps an entirely new staff of qualified professionals. Alternately, you may find it easier to hire an outside service vendor to do the work for you. An intermediate approach is to hire an eCommerce consulting team to evaluate your needs, and recommend a clear course of action before development begins.

eCommerce Best Practices

4.6.2 Advantages and Disadvantages to Service Providers

Every company has its own story, so keep in mind that there is no right and wrong on whether or not you hire a service provider. However, if you are investing in a service vendor you want to make sure that it pays off in the end. Take note of some of the advantages and disadvantages involved in service vendors.

Advantages

- A specialized, qualified staff is already in place so you do not have to create an entirely new division or search for qualified employees.
- Cost benefit—if you opt to outsource processes, the cost of the service team in their locale would be less expensive than if the team was in the U.S.
- Diffuses responsibility—this makes the service vendor accountable for the actions, and subsequent reactions, to any hang-ups or problems that may arise.
- Allows you to focus on your core business needs.
- Some service providers specialize in multiple functionalities. You may not have to go around looking at vendor after vendor—look to the partnerships you have already created because they may just provide the service you need.
- Some service vendors are great for smaller, mom & pop business because they can expose their products to a wider audience (e.g. PayPal or Google Checkout)

Disadvantages

- Your direct involvement may be limited, so be ready to relinquish control of anything.
- Managing vendors could prove to be much more difficult.
- Expect the unexpected. Bugs or disruptions can pop-up, but do not be alarmed, it should be the service vendors responsibility to handle it.
- Security issues involved with offshore outsourcing may arise.
- Someti mes offshore outsourcing will return insufficient metrics to determine the quality of service.
- Service vendors like eBay get kickbacks for every item sold.

eCommerce Best Practices

4.6.3 Offshore IT Outsourcing

Outsourcing has recently become a hot topic in terms reducing implementation costs for many IT-related projects. Certainly there are risks involved in managing team members remotely, but if managed correctly a project can be accomplished in less time and with lower expected expenditures.

There are no guarantees that offshore outsourcing is the best option for your company, so examine your reasons for outsourcing — if it is to only reduce costs, there may be other options rather than having to set-up or seek the services of an offshore branch. These may be hiring a PMO from a professional services firm and staffing the project with in-house talent. In this model, you not only benefit from a team of expert project managers, but you share the risk with a third-party contracter.

4.6.4 Managed Services

You may choose to opt for entering a relationship with a service vendor that entails greater involvement, typically in IT processes. This is called a Managed Service and it is designed to allow greater direct communication between the parties involved, as opposed to offshore outsourcing.

> **A managed service is a good opportunity to enter in a long term partnership in terms of your eCommerce functions and capabilities.**

Depending on the type of service, some partnerships with service vendors can last years for a full-cycle or long-term maintenance project, or months for short-term technical assistance, etc. In some cases, the level of involvement is particularly high because employees or consultants may need to work at the client side offices. You will be responsible for overseeing their performance but you do not have to relinquish total control of the management itself.

eCommerce Best Practices

4.7   Infrastructure Security

    4.7.1   Introduction
    4.7.2   Credit Card Security
    4.7.3   Traditional Mechanisms for Network Security & their Drawbacks
    4.7.3   Vulnerabilities Associated with Database Secuity

4.7.1   Introduction

To build a strong house, requires a solid foundation. And, for eCommerce implementations this requires a well-designed and managed support system for security. While the popular notion of security topics often focuses on configuring routers, installing firewalls, securing application servers, or providing secure application access will make applications secure, the most effective approach to network security is to layer security solutions at different points within the network. With this model, all modules work together to adopt an in-depth defense strategy to keep the network safe from malicious attacks. Each layer provides some protection from unauthorized intrusion or some specific form of malicious behavior, and, thus, the entire network is not in jeopardy when one layer has been compromised.

> **Layer security solutions at different points within the network and do not rely on a single solution.**

4.7.2   Credit Card Security

Whenever a person submits his or her credit card information on a web page, even though there is a reassuring lock icon at the bottom of the browser window, that lock icon just indicates that the credit card number is being encrypted while in transit over the Internet and decrypted on the other side. The merchant then usually stores the credit card number in an order processing database, sometimes without appropriate levels of encryption or other security measures.

During this process, a hacker has two key opportunities to get access to the sensitive information submitted by the customer:

1. While the information is being transmitted over the network
2. While the information is stored on database server

Practically every business must take action to secure their Web Applications and the Databases in order to protect the sensitive credit card data that they store, transmit, and receive.

4.7.3 Traditional Mechanisms for Network Security & Their Drawbacks

Traditional network security methods such as firewalls and intrusion detection systems are important front-line barriers to would-be attackers. However, security has many aspects and dimensions. Designing applications so that they are secure is a neglected but critical aspect of an in-depth defense approach to securing applications. This section discusses how these methods can prove to be insufficient in the deployment of a secure system.

Firewalls
The term Firewall is an interesting buzzword in the digital world. The makers of Firewall software and hardware were able to associate Firewall as a one shot solution for security concerns: "We have a firewall in place and therefore our network must be secure." However, total reliance on a firewall may provide a false sense of security. The firewall will not work alone (no matter how it is designed or implemented) as it is not a panacea. It is simply one of many tools in the toolbox for the implementation of IT security protocol.

Regardless, firewalls are essential for security and are able to prevent many unauthorized intrusions. As stated earlier, a firewall is not an effective single point of defense because of the things it cannot protect against, such as:

- Services that have to pass through the firewall.
- Attacks not made through the firewall (i.e., from inside the organization).
- Trojan horse programs.
- Completely new threats.
- Bad or non-existent policies.
- Incorrect filter setup.
- Sophisticated attacks.

- Low and Slow attacks.
- Vulnerabilities within the firewall implementation itself.

Intrusion Detection Systems

Network Intrusion Detection Systems (NIDS) are designed to monitor packet transfers over the LAN, a crucial piece in the security landscape and an excellent complement to the firewall. NIDS can provide worthwhile information about malicious network traffic and alert security personnel that someone is trying to break in, or has already succeeded in doing so.

Intrusion detection is a necessary element of network security but cannot stand alone because of its emphasis on detection. By the time corrective action is taken, damage to the network has already occurred. Further, NIDS primarily protect against attacks seen previously. Some systems attempt to identify deviations from normal activity. However, definitions of "normal" activity must be very loose to avoid generating too many false alarms, which opens up opportunities for circumventing the detection mechanisms.

As in the above case of firewalls, attackers that seek to intrude systems undetected may choose to attack the server which hosts an intrusion detector or anti-virus application to subvert the application so that it will not detect that particular attacker's activities.

4.7.4 Vulnerabilities Associated with Database Security

People generally associate security procedures with web servers and not databases. However, just because front-end security processes are in place does not mean your database is safe from security risks. The database is still highly vulnerable to hackers and information infringement.

In eCommerce, databases can contain valuable information about the company's assets. Putting that information at risk is not worth the integrity of your company. The consequence of an insecure database is more than just a PR problem. Embezzlement, identity theft and extortion are all types of security risks to a breach of database.

Because highly sensitive information is stored in many eCommerce companies' databases, you are responsible for securing your entire customer base's personal data. This includes credit card numbers, social security numbers,

financial accounts, contact information, etc. If this information gets leaked identities are compromised and can be used illegally. Without question our organization will be scrutinized for lack of security protocol, which may result in Federal legal issues.

Another risk of hackers accessing customer data is extortion. Many hackers believe companies would rather not make the security breach public, which means they would avoid the police or press involvement. If credit card data gets into the wrong hands, you are liable for people's identities.

Databases are undoubtedly a target for hackers because the information stored in them is highly sensitive. Any security breach to databases is troublesome for all parties involved. Hackers have more financial incentive to steal people's identities, than to hack just the user interface or design of a site. Database theft affects your revenues, especially in eCommerce. By protecting the data within the database, organizations can provide a final line of defense before data is compromised.

> **Vulnerabilities always exist in implementations and deployments of database services over the Internet.**

eCommerce Best Practices

4.8   Application Security
   4.8.1   Introduction
   4.8.2   OWASP
   4.8.3   SSL & HTTPS

4.8.1   Introduction

Web security issues have been flooding the media and Internet news-wires with greater frequency. Security breaches are unfortunately common these days, especially for eCommerce related businesses as hackers become more opportunistic attackers. In other words, they see system vulnerabilities and they want to exploit them. Once the infrastructure security model is in place, it is time to focus on the user-facing application as a front-line defense against malicious attackers.

4.8.2   OWASP

The Open Web Application Security Project (OWASP) is an opensource organization focused on constant application software improvement. Each year, the OWASP publishes a list of the "top 10" application security vulnerabilities. The 2007 list from www.osasp.org is documented on the following page2:

---

[2] http://www.owasp.org/index.php/Top_10_2007

eCommerce Best Practices

A1 - Cross Site Scripting	XSS flaws occur whenever an application takes user supplied data and sends it to a web browser without first validating or encoding that content. XSS allows attackers to execute script in the victim's browser which can hijack user sessions, deface web sites, possibly introduce worms, etc.
A2 - Injection Flaws	Injection flaws, particularly SQL injection, are common in web applications. Injection occurs when user-supplied data is sent to an interpreter as part of a command or query. The attacker's hostile data tricks the interpreter into executing.
A3 - Malicious File Execution	Code vulnerable to remote file inclusion (RFI) allows attackers to include hostile code and data, resulting in devastating attacks, such as total server compromise. Malicious file execution attacks affect PHP, XML and any framework which accepts filenames or files from users.
A4 - Insecure Direct Object Reference	Frequently, an application only protects sensitive functionality by preventing the display of links or URLs to unauthorized users. Attackers can use this weakness to access and perform unauthorized operations by accessing those URLs directly.
A5 - Cross Site Request Forgery (CSRF)	A CSRF attack forces a logged-on victim's browser to send a pre-authenticated request to a vulnerable web application, which then forces the victim's browser to perform a hostile action to the benefit of the attacker. CSRF can be as powerful as the web application that it attacks.
A6 - Information Leakage and Improper Error Handling	Applications can unintentionally leak information about their configuration, internal workings, or violate privacy through a variety of application problems. Attackers use this weakness to steal sensitive data, or conduct more serious attacks.
A7 - Broken Authentication and Session Management	Account credentials and session tokens are often not properly protected. Attackers compromise passwords, keys, or authentication tokens to assume other users' identities.
A8 - Insecure Cryptographic Storage	Web applications rarely use cryptographic functions properly to protect data and credentials. Attackers use weakly protected data to conduct identity theft and other crimes, such as credit card fraud.
A9 - Insecure Communications	Applications frequently fail to encrypt network traffic when it is necessary to protect sensitive communications.

4.8.3 SSL & HTTPS

From an eCommerce application perspective, it is critical for customer transactions be controlled in a secured environment. To accomplish this, a security layer is introduced into 3-Tier or N-Tier architectures called the Secure Socket Layer, or SSL. It is at here where the encryption and decryption of private data occurs. By isolating this information, it can become removed from the application layer and database layers and thus reduces its ability to be accessed by unauthorized users.

From the Customer's perspective, the passing of private information to the eCommerce application is seamless with one slight difference. Insecure websites are contacted through HTTP, or Hyper Text Transfer Protocol, while a secure website requires an HTTPS request. It is this nomenclature that acknowledges to the user that they are connecting to a secure site and have permission to pass private data via SSL.

The implementation of the secure layer requires an authentication program to complete the secure "handshake" between user and application. This is known as an SSL Certificate, and it is generally a few hundred dollars in cost. Marginal when compared with imagining the possible problems and risks of exposing your buyer's credit card data. Some popular SSL Certificate venders include the following:

- Verisign: www.VeriSign.com
- Wildcard: www.DigiCert.com
- GoDaddy: www.GoDaddySSL.com

4.9 Infrastructure

- 4.9.1 Introduction
- 4.9.2 Infrastructural Components
- 4.9.3 Hosting
- 4.9.4 Applications
- 4.9.5 Platforms
- 4.9.6 Monitoring and Support

4.9.1 Introduction

"eCommerce" — the word itself should direct your thoughts to websites, web shopping carts, digital product catalogs and online payments. Out of these things, the primary focus is likely on the website and how it looks and feels. The look, feel and usability of the site is, of course, a high priority and it would not be a good idea to ignore the aesthetic appeal of the site, but in the long run another major aspect of successful eCommerce implementation is the infrastructure that supports the site. Here are a few examples on why this is true:

An amazingly interactive Flash based site would mean nothing if the user has to wait for a long time before the site becomes usable.
A highly dynamic site based on latest technology would not be so attractive if it is not available to the user.
However famous or attractive the eCommerce site, if it does not provide valid security and privacy it would not entice the regular web-surfer to make use of it.

These aspects are the major driving force behind the success or failure of an eCommerce site and to address such concerns, it is a fundamental requisite to have a robust infrastructure that supports it. eCommerce is not just about a pretty looking façade, but it is also what goes on behind the scene; it is the eBusiness which caters to the success of the site. This eBusiness may include different facets of commerce like Electronic Funds Transfer (EFT), Supply Chain Management, Online marketing, Online Transaction Processing (OLTP), Automated Inventory Management System, Electronic Data Interchange (EDI), specifically Electronic Invoicing and Electronic Purchase Order Generation.

Such a complex business process requires highly scalable and secure infrastructure that can withstand the forces or user load, natural disasters, and of course human disasters too.

4.9.2 Infrastructural Components

Infrastructural components can segregate such into three main categories:

Hardware
Hardware requirements to support the eCommerce infrastructure can encompass various items like Hardware Web Servers, Multiprocessor Server Machines, Server Clusters, Mainframe Servers, VOIP terminals, Voice Response Systems.

Network
The network consists of gigabit Ethernet Switches, Routers and Firewalls along with cabled or wireless network setup for creating a LAN (Local Area Network) or a WAN (Wide Area Network). Creating a server cluster which suits the need of your business is not child's play. It requires accurate information about the business scope and growth and as well as keeping in mind the Cost of Ownership and Cost of Maintenance

Software
Software components are required to run the hardware efficiently, run the business of the relevant corporation and manage the network. The application software will be discussed further in the chapter, but before it is explained in greater detail, take a look at a few typical infrastructure scenarios.

eCommerce Best Practices

Infrastructure Scenarios:

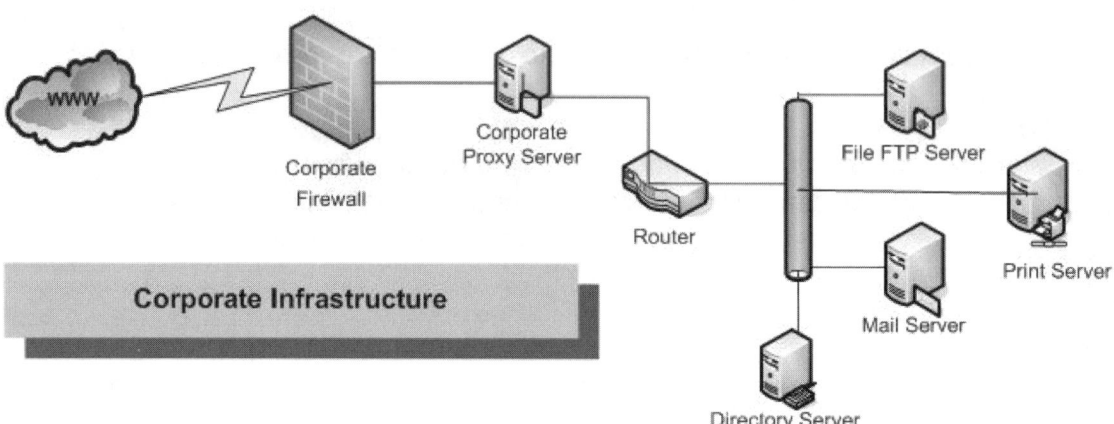

Figure 1. A typical corporate infrastructure.

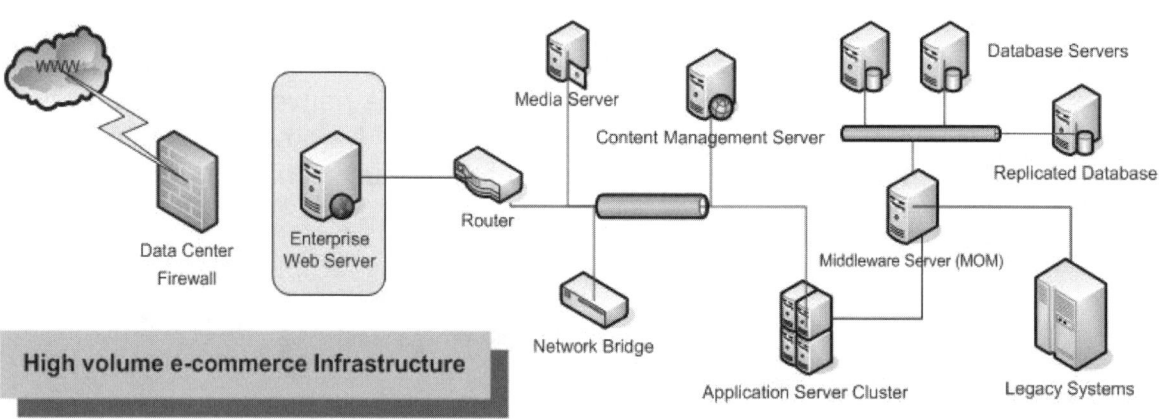

Figure 2. A typical eCommerce infrastructure

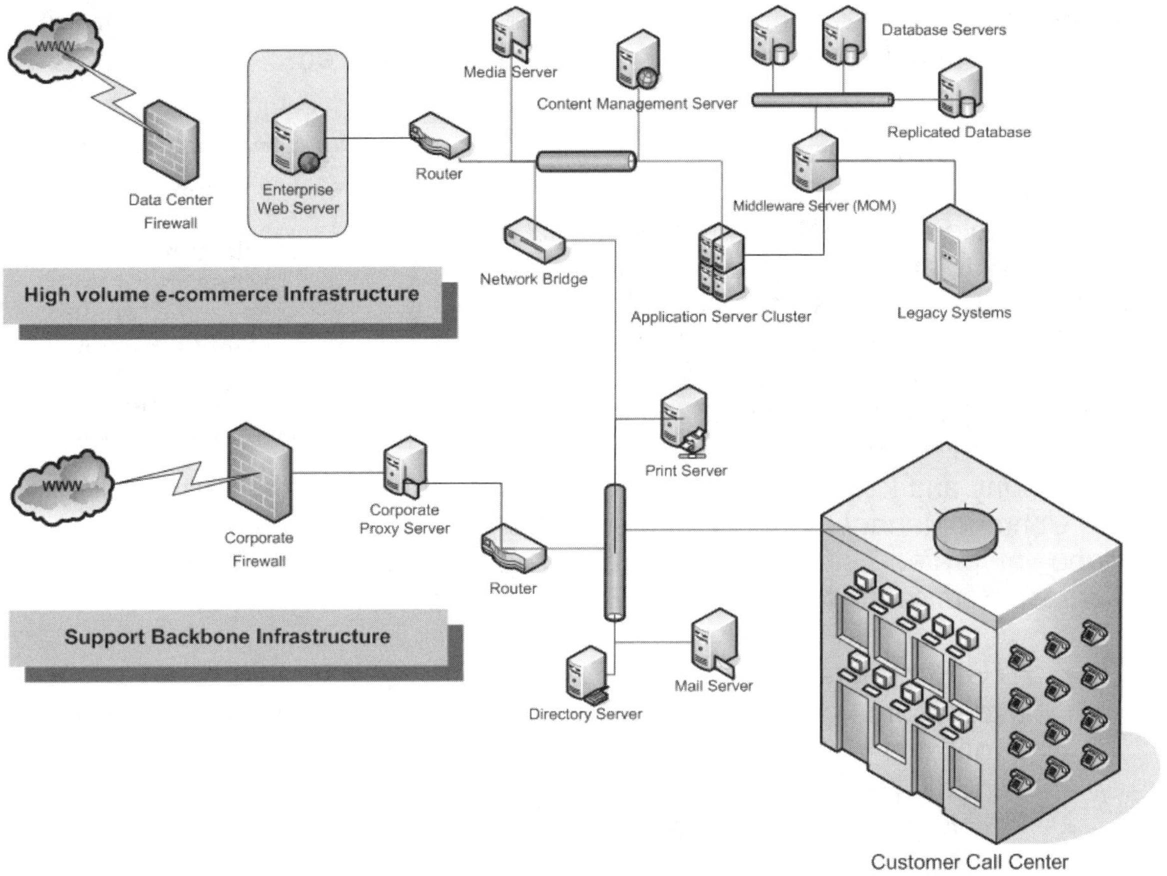

eCommerce Best Practices

Figure 3. Typical eCommerce infrastructure backed by call center

4.9.3 Hosting

These infrastructures can be typically be found in large corporate houses or hosting facilities that provide eCommerce infrastructure for hosting an eCommerce site.

eCommerce hosting is a service that allows organizations to provide their own eCommerce websites for the world to use. These hosting companies provide physical space, as well as their server resources, to their customers who are corporate houses. Such a facility is called a Data Center. Many hosting companies provide collocation of the servers wherein the servers are not actually owned by the hosting company, but they just provide the space, connectivity and environment for servers.

eCommerce Best Practices

There are various components to manage hosting as shown by the diagram; these components can be broadly categorized as followed:

4.9.4 Applications

To assemble a complete eCommerce package it is necessary to include various components at application level. These components provide the functionality that is necessary for a smooth execution of business processes. Applications are the heart of machines, and for a healthy eCommerce site many applications need to work in harmony and keep the site alive and kicking. You can enlist such applications like Content Management Tools, Media Tools serving images or streaming video, Integrated Search Engines, Shopping Carts, Payment Gateways, Address Verification Services, and Email Servers:

> **It is important to assess the applications that are necessary for your site, and to ensure that they work cohesively with one another.**

- ATG Commerce Suite (http://www.atg.com)
- Cybersource (http://www.cybersource.com) [Payment Gateway, Address Verification]
- VeriSign Payflow (http://www.verisign.com) [Payment Gateway]
- Paypal (http://www.paypal.com) [Payment Gateway]
- Authorize.net [Payment Gateway]
- Zen Cart [Shopping cart]
- BakeSale [Shopping Cart]

4.9.5 Platforms

Platforms are the foundation of eCommerce sites. They mainly consist of a stable Server Operating system which is multi-processor capable and can scale horizontally. This Server Operating system forms the robust pillar on which the whole infrastructure web can stand. These Operating System environments are generally geared towards Enterprise capabilities.

At the top of the Operating system are the Application Server and the Web Server which provide runtime environments for the application to run and respond to internet requests. Finally, there are the integration platforms which provide capabilities to connect to Legacy Systems; such platforms are mainly Message Oriented Middleware or (MOM) servers:

eCommerce Best Practices

- AIX , Unix , Solaris [Multiprocessor OS]
- BEA Weblogic Application Server (http://www.bea.com)
- IBM Websphere Application Server (http://www.ibm.com)
- IBM MQSeries (MOM)
- TIBCO Rendezvous

4.9.6 Monitoring and Support

Often with such a large investment in building a successful infrastructure, it is necessary to keep a close watch on the happenings of the website. This monitoring facilitates the hunting of information that can be very crucial to the business. Moreover, the monitoring provides a health check for the system as well as the growth of the site.

Monitoring it is typically a test conducted on the website from inside and outside. Apart from tests, monitoring also consists of tracking user activities and how the system behaves to the stimulus provided by user actions. Website monitoring is also used to ensure the "availability" and usually hosting companies assure availability higher than 99%.

> **You should monitor how the system reacts when users performs certain actions.**

There are numerous tools that monitor a site at different application and hardware levels. Various network tools provide deep insight into network bottlenecks, usage and traffic. They help in analyzing the throughput of the website and the data transfer rates. There is help in gathering crucial information for load testing of the website:

- Aurora (http://www.zurich.ibm.com/aurora/)
- Netflow (http://www.cisco.com/)

For application monitoring and support there are dedicated services that offer this functionality. This includes monitoring the health of the Application Server, Database servers and mail servers, and proper execution of various protocols like HTTP and HTTPS:

- WebwatchBot (http://www.exclamationsoft.com/)
- AlertSite (http://www.alertsite.com/)
- Gomez (http://www.gomez.com/)

The list can grow vast as more and more players develop new offers and unique selling points, but the main aspect to keep in mind is the seamless integration of disparate components. A lot of the shelf software suites are available which provide end to end services, and choosing the best option for the business is the key to success.

eCommerce Best Practices

4.10 Analytics / Business Intelligence

4.10.1   Introduction
4.10.2   Web Analytics
4.10.3   Data Analytics
4.10.4   Conversion Funnel
4.10.5   A/B
4.10.6   KPI
4.10.7   Tools Category

4.10.1   Introduction

Analytics and BI (Business Intelligence) are important cornerstones for building a strong eCommerce platform and business. They entail quantitative techniques to measure how certain indicators are doing and suggest some actionable items.

While analytics takes a micro-level perspective in the form of web analytics (assessing in web site level), or data analytics (assessing in data level), etc., BI takes a higher or macro level perspective and deals at levels like business operations, strategy, etc.

> **The techniques leverage and strengthen expert knowledge, judgment and wisdom but they never replace or minimize need of them.**

Because business operations and strategy are essentially supplementary to each other in a modern electronic based commerce and because of the web platform with associated data, the quantitative techniques of Analytics and BI, together, are able to provide more complete information that an eCommerce company would need in order to succeed in a highly dynamic and competitive environment.

This chapter will present some pertinent matters related to these two important techniques and will try to suggest some recommended practices for their usage.

eCommerce Best Practices

4.10.2 Web Analytics

Web Analytics make an attempt to measure the effectiveness of web sites and/or particular pages of sites in terms of their ability or propensity to affect user behavior.

Several techniques have been in use including analysis of logs, capturing each instance of page visits, etc.

> **Implement and use an analytics technique to measure the effectiveness of your website.**

The suitability of a particular technique is dependent on the state of the operating environment and some technological issues concerning how pages are generated and rendered. The user's interaction with a particular page may not always necessitate a generation of pages in the server side when intermediary technologies like proxies or local caching have been used.

As a general rule of thumb, analysis of logs is simpler to implement and may meet requirements for a majority of the instances. Moreover, the majority of system installations already have a logging mechanism in place and some even have very comprehensive logging systems. In those instances, web analytics with analysis of logs are an extension of previously built logging systems.

In the situation where proxy servers are involved or data is cached at a local level, the analysis of log files in the server may not always reflect the actual user's interactions with the system. There are techniques to capture each instance of a page visit that provides deeper and more accurate matrices, but this requires major code rework, more complicated system infrastructure, etc.

4.10.3 Data Analytics

Data Analytics concern itself with analyzing data including the data entered by user/ visitor in order to make some useful inferences about behaviors, patterns, and effectiveness of a particular aspect of business. It entails various effective modeling techniques and can provide results that can be important for strategic, tactical, or operational decision making.

> **Use data analytics techniques to turn raw database into effective knowledgebase.**

eCommerce Best Practices

Techniques such as data warehousing and data mining are well known in the industry. The two are instances of data analytics techniques. Other techniques might include the predictive modeling technique.

The knowledge derived from source data should, however, be validated and the mechanism should be set up so that, if possible, the inference engine can continuously learn more from the data that it analyzes. There are various tools available for providing more effective data modeling and making better inferences out of the models. Selection of a particular tool should be guided by your scope or requirements, investment cap, domain specific issues, etc.

4.10.4 Conversion Funnel

A conversion funnel is an abstraction of the visitor's purchase process in an eCommerce context. It is essentially the path that takes place from first visiting site to finally purchasing the item. There are various factors that contribute to a visitor's decision making as he or she goes through various small steps involved in the process. The key is how the visitor can be led to continue in the steps rather than go to other parts of the site or leave the site. Generally, the conversion funnel is broken down into steps like browsing items, checking details, putting in carts, providing personal information, and making a purchase, etc.

> The conversion funnel-human purchase process can be broken down into little steps. It is important that the steps be understood and influenced from human behavioral perspective.

Many people reasonably argue that the so called linear approach to a conversion funnel breakdown does not represent the reality. There are fundamentally various uncertainties and non-determinisms that are involved in each step of the process and rather than trying to understand the funnel mechanically, it is important to understand and influence the funnel (the steps in the process) with a human behavioral perspective.

4.10.5 Actual Behavior (A/B) Testing

Actual behavior (A/B) testing is one of the most widely used techniques to quantitatively assess how users behave for a given situation or respond to a given product or product design. In this method, a specific segment of user is given a new or different choice in design or specific kind of service delivery, and an assessment is made as to how that specific user segment behaves differently than other segments in terms of some quantifiable parameters.

This technique is good when the assessment objective is quantifiable and specific. This technique, however, has limitations and cannot properly assess intangible, broad, and long-term objectives.

This technique is widely used and is recommended because of its simplicity, low cost and it being a historically effective technique.

For a long term broad objectives assessment, however, techniques involving human behavioral perspective need to be involved. For still more comprehensive assessment, both of the techniques may be used in combination.

4.10.6   Key Performance Indicators

Key performance indicators (KPIs) are used in Business intelligence to capture values of parameters that have an impact on the direction or speed the business has to take in order meet the set business objectives.

What constitutes KPIs is determined by the nature of business, set objectives, prevailing environment and so on, hence it is not possible to generalize and express in a way that is relevant in any context. The important point, however, is to make them quantifiable, relevant, and achievable.

> **Key performance indicators (KPIs) that are important for business intelligence should be quantifiable, specific, and achievable.**

The assessment of KPIs is generally intended to lead to some kind of action—be it corrective, preventative, or just making a recording. The KPIs are categorized in a suitable manner that is compatible to organizational context, business domain and defined objectives.

4.10.7   Tools Category

There are wide varieties of BI tools available and there are offerings from numerous vendors for the same category.

eCommerce Best Practices

While it is not possible and may not be appropriate to make an exhaustive search and treatment for each offering, it may be helpful categorizing them suitably so that it will create a better conceptual understanding as to how each tool fits in the big picture. The understanding will foster better selection and tools usage.

Business intelligence tools fall on one of the following categories:

Reporting tools that connect to various data sources and provide related information derived from data in the form of reports, charts, etc.
Online Analytical Process (OLAP) engines that can slice and dice huge chunks of data from different perspectives and can generate some useful information out of the data.
Data mining that goes deeper into huge chunks of data like a data warehouse and makes an attempt to find some trends or patterns in data and come up with some knowledge about a particular aspect of business.
Business Performance Management (BPM) tools that not only derive the information or knowledge, as did the other tools categories explained above, but goes one step further and tries to use the results to improve the performance of your business—thus creating an intelligent feedback system.

The order of categories shown above is in approximate order of complexity, cost and capability. Business needs, as usual, drive it all.

As a rule of thumb, use reporting tools or OLAP for generating information which can be further processed by a human expert to derive some knowledge. For a smaller set of databases and simpler needs, reporting engines may be adequate. For large scale databases and for more sophisticated analysis, OLAP techniques are better options.

For systems to generate knowledge by diving deep into large set of data, Data Mining is the technology you want. If you want to set an automated system that not only generates business knowledge but goes beyond that to make an attempt at using the knowledge for business improvements (an intelligent feedback control system), BPM is the option to choose.

eCommerce Best Practices

4.11 Personalization

4.11.1 Introduction
4.11.2 Why Personalize?
4.11.3 Elements of Personalization
4.11.4 Building User Profile
4.11.5 Content Delivery
4.11.6 Tips for Personalization
4.11.7 Personalization—A Myth?

4.11.1 Introduction

Personalization is only as good as the relationship you earn with the consumer. Knowledge is power, and the more you know about the person buying your products the more you will gain in the long-run. For example:

You walk into a crowded bar when the game is on, the music is blaring, a large group of people are waiting to get their orders in and the bartender immediately caches your eye. You nod. Within seconds you get your "Double Black Label scotch on rocks". Not a word exchanged. The bartender gets a fat tip, you go on your way, and yet he knows you will be visiting him again sometime soon.

The scenario describes the essence of personalization. In the world of eCommerce, the bartender represents the web site, you represent a user on the web site, and your scotch on the rocks is the information you're customer is looking for. Personalization, in brief, is the matching of information to the user's interests.

> **Personalization is the matching of information to the user's interests.**

When you buy a book through an online retailer, like amazon.com, you often see recommendations displayed to you; "People who bought this also bought." This is often a quoted example of personalization. You will see more examples of personalization in this chapter.

4.11.2 Why Personalize?

In the competitive world of eCommerce, repeat buyers are worth their weight in gold. What do web sites do to get their users to re-visit them and spend their hard-earned paychecks? They personalize.

You might not realize it, but your shopping experience online is often personalized based on the things you purchase. Take, for example, buying a new television — the likelihood of you walking away, or in this case "clicking" away, with only a TV is slim. Your intentions might be to only purchase a big screen HD plasma, but after adding it to the cart, you are taken to a page for cables evaluating various cables (the component, S-Video, HDMI).

After realizing you need the cables in order to hook up your TV and after adding them to the cart, you are taken to a page on TV stands. You were not planning on wall mounting it, so their recommendations for a TV stand that conceals the wires seemed like a good alternative. After it is all said and done, you end up purchasing things you did not even know you needed, and then you receive a discount coupon for your next purchase from their site, which will undoubtedly be used on something—who can pass up a coupon?

The web site had personalized shopping experience by targeting the right products to you at the right time. Their recommendations were filtered and did not overwhelm you with mindless products. They had the right number of recommendations, as well as the right types of products, making it more relevant to you as a consumer. When the time comes to buy a new digital camera, you will most likely return to their site, armed with a coupon, and this time the process will be even quicker because your billing and shipping information has already been registered with them.

What the web site had achieved was this—they attracted you to their site by providing an enjoyable shopping experience. They made you return to their site by luring you with discounts and topped it by improving your shopping experience. Your repeated visits to the site contributed in small measure to their increase in sales. Factor in the thousands of other users who might have enjoyed a similar experience, and chances are it will experience continued success.

4.11.3 Elements of Personalization

As we have seen earlier, personalization is matching information or content to users' interests. A typical eCommerce web site houses many user interests that fall under a broad spectrum. Capturing as much information as possible about each user is essential to the success of personalization. The information captured is used to build a profile for each user.

eCommerce Best Practices

eCommerce web sites have a lot of content and it is critical that the content is clearly defined and organized. This makes mapping of the content to user's interests a lot easier, however varied their interests might be.

To summarize the key elements of personalization are the user profile and content.

4.11.4 Building User Profiles

How do web sites go about collecting information about what their users are interested in buying? If you are reading this book, chances are rather high that you might have completed at least one online registration form. Visitors are asked to answer a number of carefully drafted questions. Answers to these questions are used by web sites to determine how best to customize their shopping experience. This form of information gathering is called explicit profiling. The advantage of this approach is that the user is forthcoming with the information. No guesses required. The disadvantage is that in this day and age, people are over-cautious about the information they provide to online vendors. Often enough, responses to some of the questions may be false.

> **Have users complete a quick online registration form to acquire valuable information in the initial personalization process.**

Explicit Profiling and the Anonymous User

If web sites personalize using only the explicit profile, then personalization might not be effective and, infinitely worse, might backfire on them and they might start losing users. Explicit profiles are created only for users willing to register with the web site. This effectively rules out personalized shopping experience for users not registered with the site (known as Anonymous users). The pitfall to this is that there is a major disparity between the experience of a registered user and an anonymous user. Anonymous users should be treated as prospective registrants. A good shopping experience for anonymous users might lure them to register with the site. An explicit profile will then be created for them.

Click-stream Data

Another approach to determine user interest is capturing click-stream data as the user navigates through the site. Click-stream is information on what pages a user viewed, what searches were performed and which links were clicked. This information can be captured for both anonymous and registered users. In

the case of anonymous users, the click-stream can be used to perform some amount of personalization. In the case of registered users, click-stream data can be used to fine-tune the user's interests. Click-stream data is an implicit way of capturing user information.

The Power of Two

In the context of determining user interests, these two approaches are not mutually exclusive. They go hand-in-hand and complement each other. When used effectively these prove to be potent weapons. In explicit profile creation, the online registration form is not just any odd form to capture information. The questions presented to the user should be limited enough in order to minimize the time required to complete the registration process while being succinct enough to gather as much information as possible. A lot of planning is required in drafting an effective registration form. In the implicit click-stream approach, within a few minutes of turning on click-stream data capture, the amount of data captured will be huge. Careful analysis of terabytes of data is required to glean knowledge from the captured data. Poor analysis of click-stream almost always leads to poor personalization.

4.11.5   Content Delivery

Even with the most accurate user profile and the most advanced mining tools working on click-stream data, poor quality content delivered to users defeat personalization efforts.

Content is Vital

In the HD-TV scenario described earlier, the web site targeted delivery of cables required for the TV along with an analysis of the cable technology. First time HD buyers may not be aware of different cable options. If a web site simply targeted the various options, then users might be overwhelmed and in their quest to understand the technology may navigate away from the web site. This leads to loss of revenue to the web site, and in some cases may lead to shopping cart abandonment. The user originally had the TV in the cart and in the quest for information on cables, navigates away from the web site and abandons the cart.

Rule-Based

There are a couple of approaches to content delivery—rule-based and filtering. In the rule-based approach, business users draft rules for

> **Segmenting users by different attributes or classifications can maximize purchasing potential when sending content.**

content delivery. An example would be delivering product Y to users who have added product X to their cart. This is similar to stores placing eggs in the same aisle as milk. To the consumer who comes to the store with a shopping list it would not matter if the eggs were in the same aisle as milk or if it was in the personal hygiene section. But to the consumer who comes to the store to get some milk, noticing the eggs next to the milk, might prompt them to buy eggs.

Filtering

Filtering works on user groups. Web site users are often classified and slotted into different groups. Content is then matched to each of these groups. These groups could be as broad as male/female or based on more complex grouping of user attributes. Besides users, content too can be classified. Content is often tagged with meta tags describing the document. User profile attributes may be matched to document classes and delivered to users.

It is often beneficial to cache content for better performance. In the rule-based example, every user who buys product X is going to be displayed product Y. When the first user accesses product X, content for product Y can be cached and displayed to other users from the cache.

4.11.6   Tips for Personalization

Building user profiles, analyzing click-stream data and matching content to user's interests can be done in numerous ways. Every site would capture different information about their users. Similar sites catering to the same user pool can still have different user attributes. Every site can choose a different tool for analyzing click stream data. Sites may choose not to analyze click-stream data. Rule-based and filtered content deliveries are two approaches to delivering content. Sites can come up with their own custom approach to delivering content. Broadly speaking, the following are some tips for personalization:

- A simple greeting when the user logs in gives the user some confidence that the site is working towards serving him better.
- Carefully analysis is required for placement of targeted content. Measuring user responses to various strategically placed contents can be measured and used in determining placement of future content.
- Often the look and feel of the targeted content drives users away from the content. Flashy and risqué promotions have a negative effect on users. Of course, the class of audience your site caters to determines it.

- Prompting users to provide information frequently should be avoided whenever possible. For example, when the user checks out of the shopping cart and provides all required credit card, billing and shipping information and completes the transaction and starts afresh, if the user is prompted with the same information during the second check out might lead to some users abandoning their carts.
- Sites usually partner with other sites. Prompting users to login each time they navigate from one partner site to another should be avoided.
- Overwhelming users with promotions can often be seen as badgering by users. The thin line between enhancing the shopping experience and being a pushy sales rep should never be crossed.

4.11.7 Personalization — A Myth?

Jupiter Research recently published a report questioning the effectiveness of personalization. They show numbers in the report indicating that a small number of vendors have benefited greatly from personalization while a large number have not seen a good return on their investment. They go on to add that a flexible navigation and a good search can replace personalization in most cases.

eCommerce as discussed in this book is the harmonious union of a number of different elements. Each element serves to complement the rest. For example in the milk and eggs scenario, the consumer can search for eggs and find them in the personal hygiene aisle. Or the store can provide signs all over the stores to direct users to eggs. It might cater to certain users and this might turn off certain other users.

Software that aids in personalization is not simple enough for vendors to plug-in to their application and expects results immediately. Detailed design, planning and effective content creation are essential in delivering a truly effective personalized user experience.

4.12 Search

4.12.1 Introduction
4.12.2 Faceted Search
4.12.3 Consumer Insight
4.12.4 Discovering New Items
4.12.5 Search UI Considerations

4.12.1 Introduction

Earlier Search was mentioned in the context of Search Engine Optimization and Marketing. This is extremely important because it effectively leads people to your site that, otherwise, may not have found it. Search engines are the primary mode of entry in to the World Wide Web for most users today and any site that ignores the importance of its page ranking will surely suffer. But what happens after potential customers are driven to your site—what then? Approximately, 40 % of visitors entering a website's homepage will use the site's search mechanism rather than rely on its navigation tools. This is why you cannot ignore the power of your own site's search capability, not just the search engine optimization.

Search is a standard feature in any eCommerce website today; however, the quality and ability of a search mechanism can vary from site to site. Many times when visitors type in keywords about items of interest to them, they either get a results page with a range of different products or very few products at all. This may not be because you do not have what they were looking for, but because your search capability was not able to match what they were looking for.

4.12.2 Faceted Search

One of the most frustrating things for online shoppers on a website is when they know what it is they are searching for and not being able to find it within the first couple search attempts. This is inexcusable and absolutely avoidable if you have a sound search mechanism in place, meaning that your product catalog repository is categorized in the

> Faceted search is an effective way to drill-down the customers' search results by making them relevant to his or her specifications.

eCommerce Best Practices

most optimal way possible. When visitors use the search mechanism you want to make sure you are filtering out irrelevant products and displaying the most relevant ones.

One of the most effective features to a search function is being about to refine and filter products based on specific attributes. After the result of a search are shown, the visitor can further specify what it is they are looking for based on various categories that the product catalog is divided into. This drill-down effect is known as faceted search—where consumers can define their requirements and are only shown the products that meet them. This gives consumers more time to shop for what they want instead of wasting their time on finding an array of possible product matches. Common faceted search categories include:

- Price
- Color
- Size
- Brand
- Rating
- Style
- Function

If a consumer has a certain budget or wants a certain color, this is a way that makes their online shopping experience more enjoyable and less time consuming. Clutter is generally considered an undesirable trait, and that certainly applies to an eCommerce site from a consumer's perspective as well as a retailer's perspective. That does not mean you should not cross-sell or up-sell, so long as you can decipher what products are most relevant or thoughtful. Being able to "trim the fat" will benefit you and your customers in the long the run.

4.12.3 Consumer Insight

Search provides a means of acquiring customer intelligence and knowledge. The keywords and products that customers enter into the search box explicitly tell you what they are interested in, concerned with, or curious about. This is a great source for personalization because you learn how

> **The keywords and terms typed in the search box are extremely useful in discovering customer interest and behavior.**

each customer arrives at the product he or she was searching for—in relation to the terms or word pairings used in the search box, price range, brands of interest, etc.

If you just look at the search terms used, you know what the customer wants or at least is interested in. That information, along with previous search terms used in past online shopping endeavors, is helpful in forming a purchase pattern profile for each customer and learning about the types of products they have an affinity for. Behavior can be mapped and predicted over time, so it is important that you make as many repeat buyers out of every visitor to your site.

4.12.4   Discovering New Items

With organized and categorized search results pages, customers are more likely to find other items of interest in addition to the item they had initially searched for. Even though the number of listed products may decrease from organizing the search results, the number of relevant products will have increased, making the likelihood of add-on products increase as well.

Say a customer goes to an eCommerce site with a particular product in mind and looks for it through search. He or she types in the keyword "women's tennis racquet" into the search box and hits enter. A good search results page would filter out men's products, and dwindle it down to tennis related products. A smart search results page would display other women's tennis products, i.e. tennis shoes, wristbands, strings, skirts, etc. This makes sense because the shopper has tennis on his or her mind, and would be more vulnerable and open to purchasing tennis related items. New items or products can be discovered on these search results pages, and even if the shopper does not buy any of the products, one or two might peak some interest at a later date, thus, causing the shopper to enter another search for them.

> **Customers can find other relevant products through an organized and categorized search function because you know what types of products they are interested in at that moment.**

A faceted search section can drill-down the relevant products even more by specifying the brand (Wilson, Prince, Head, etc.), price range, or possibly even grip size.

eCommerce Best Practices

4.12.5 Search UI Considerations

Search UI is getting more and more complicated each day. Users expect search pages to not only serve up relevant search results but they also want to experience new and improved ways of mining through the search results to get to the results they need. Widgets are advancements in UI methods that can make a shopping experience faster and more enjoyable.

A good example of this is the travel website Kayak.com and its use of sliders. Kayak.com uses sliders on the left navigation panel to let users adjust the price ranges on the returned results. It is heavily dependent on AJAX but is an extremely effective manner of improving usability. All the user has to do is slide the arrows to fit the price range that is in their budget, or the times they prefer to depart or arrive at their desired destination. From those stipulations, a search results page will appear that displays the available options that meet the requirements.

> **Look to widgets to enhance the user interface and experience. They can make shopping easier, faster, and more interesting.**

Another popular widget is a drag-and-drop feature that allows users to move a selection or a grouped selection by clicking and holding down items while moving it around the page, into the shopping cart, wish list, or various other site locations. There are also tracking widgets like eBay's that notify you if you are winning or losing a bid, following certain items, or selling items. New widgets can be created every day, so be open to the new forms of UI methods that can potentially increase conversion rates.

eCommerce Best Practices

4.13 Chat

4.13.1 Introduction
4.13.2 Desired Functionality
4.13.3 Pros and Cons
4.13.4 Desired Features
4.13.5 Third Party Tools
4.13.6 Current State of Things

4.13.1 Introduction

Chat is a real-time conversation between two or more users across the network. The chat is actually a text chat which takes place between the involved parties. The includes popular Instant Messaging clients such as Yahoo Messenger, MSN Messenger, AIM and Google Talk as well as public chat rooms as in Internet Relay Chat (IRC) servers and Yahoo Chat.

However, in the context of eCommerce websites the chatting is a reference to an interaction between the customer and a Customer Service Representative (CSR). This chat is typically initiated by the customer through the web browser to query about an aspect of the website and/or products. Certain chat products have a feature that can detect certain customer patterns (such as vacillating between viewing two competing products) and initiating a conversation with them; much like a sales associate offering assistance to customers on the floor of a brick and mortar store.

Instant Chat Support augments existing methods of contacting CSR such as emailing and phone support. However, Instant Chat Support would automatically submit the customer's personal information, session information and contextual information about the last viewed web page to the CSR, enabling them to better serve the customer.

> **Live Chat Support software should be customizable and should have multiple functionalities.**

4.13.2 Desired Functionality

The Live Chat Support software should provide a significant number of functionalities beyond the basic ability to perform chat with a customer. The CSR should be able to chat with multiple customers and have the ability to easily switch between them. The software should be customizable to see how the customer proceeds after talking to a CSR. And it should work seamlessly through the web browser without having to install anything on the client machine.

4.13.3 Pros and Cons

Using Chat/Live Chat Support in an eCommerce website offers a number of advantages over send email queries or talking to a CSR over the phone. However this functionality does come at a cost as well. Listed below are some of the advantages and possible disadvantages/pitfalls of using Live Chat Support.

Advantages

Instant Customer Help: The customer does not have to wait for an email response or navigate through a complicated phone menu to reach a CSR.

CSR Efficiency: One CSR can handle multiple customers at the same time by alternating between chat windows.

Page Sharing: The CSR can push pages to the customer's browser enabling them to quickly locate the desired functionality.

Reduce Shopping Cart Abandonment: Live Chat Support allows customers to have instantaneous access to help at critical points in the shopping process thereby eliminating abandoned shopping carts.

Improve Customer Loyalty: A good Chat service process can increase customer loyalty by providing for instantaneous help as opposed to waiting for 24 hours for a response to an email.

Reduce Costs: Over the long term Live Chat Services would reduce the number of calls to CSR by increasing the number of queries to online CSRs.

Disadvantages:

Comfort Level: Not all customers are comfortable talking to a disembodied text. Also a lot of people would not be as proficient in typing and would face additional frustration in getting help.

eCommerce Best Practices

Investment: Additional investment into human capital might be required if the Instant Chat Support is not out-sourced. New scripts would need to be created with the CSR in order to handle different customer scenarios. If the chat CSRs are not outsourced, then they would be require training on the chat product.

New Analysis Tools: To find the effectiveness on how the Chat service is helping sales, new tracking and reporting must be implementing to measure its effectiveness.

CSR matters more: The chat service is only as effective as the CSR's abilities. If the CSR keep the customers waiting, there will be a tendency to have disgruntled people who might not come back to the site.

Longer CSR Calls: Chat sessions will tend to last longer than the actual phone conversations, as typing takes a longer time and also because the CSR might be helping out more than one person at a time.

Outsourced Knowledge: If the chat CSR resource is outsourced out, the CSR might not know of all the functionalities or the products offered at the site and might lead to frustrated customers.

Integration: Some level of development effort would be required if the chat session is to be associated with the user or if the CSR needs to look up the customer's account information.

4.13.4   Desired Features

When looking for implementing or making an investment into a 3rd party tool for a Live Chat Support program there are a number of functionalities that should be considered. These are listed below (in no particular order):

Scripts /Canned Greetings: The software should allow for inserting pre-defined text into a chat conversation. These could be answers to typical questions, canned greetings or other frequently occurring responses such as assuring the customer that the CSR is still there and looking up information.

Ease of multi-tasking usage: The software should allow the CSR to easily navigate between different chat conversations.

> Live Chat allows you to easily monitor the content during conversations, common key words, and the extent of the conversation.

eCommerce Best Practices

Monitoring CSR: Chat software should allow for the manager to evaluate the effectiveness of each CSR interaction. Measurements include the number of daily conversations, length of conversations, and the frequency of certain key words that might be used by the customer (such as "no help").

Logging chats: All chats would need to be logged into a central database, for either the manager to view on performance of a CSR or for CSR to email a customer on request.

Tracking user activity during and after chat: Tracking a user's activity helps in determining the effectiveness and helpfulness of the chat service or a particular CSR.

Spell Checker: An on the fly spell checker would help portray a more professional CSR to the customer.

Preview text before customer sends message: Some chat software allows for sending the customer text before they actually send it. This allows the CSR to proactively searching for an answer before the customer has made a request.

Chat transfer to another operator: Just as phone CSRs can transfer a request to the right department or supervisor, a CSR on chat should be provided the same functionality.

Typing indicator: This indicates to the CSR and/or customer that a response is being typed. This allows you to see if either party is active or not, especially in the case when one of the participant is a slow typist.

Email chat transcripts: Emailing the chat transcript on customer request, allows for the customer to have a convenient reference to their queries.

Share webpages: The CSR should be able to push pages to a user to guide them to appropriate functionalities on the website.

Co-browsing: The CSR should have the ability to see what the user is browsing and to guide them properly if allowed by the customer.

Customer Survey: If enabled the customer can be presented with a survey about their experience with the CSR.

eCommerce Best Practices

SSL Secure Data Transfer: This would be required if the customer would give personal information such as credit card information over the net to prevent people from intercepting this information.

Block IP Addresses: To prevent abusive customers from wasting CSR's time, the CSR should have the ability to block certain users or IP addresses (logging the appropriate reason).

4.13.5   Third Party Tools

If you determine that developing an in-house chat solution isn't your best option, there are a wide variety of partners to choose from. Below is a list of some third-party tools that provide Live Chat Support for websites:

- Boldchat (http://www.boldchat.com/)
- eStara (http://www.estara.com/)
- Intelli-Chat (http://www.intelli-chat.net/)
- Live 2 Support (http://www.live2support.com/)
- Live Chat Now (http://www.livechatnow.com/)
- Live Person Pro (http://www.liveperson.com/)
- Live Site Manager (http://www.livesitemanager.com/)
- Livehelper (http://www.livehelper.com/)
- PHP Live (http://www.phplivesupport.com/)
- Site Chat Deluxe (http://www.sitechatter.com/)
- Velaro (http://www.velaro.com/)

4.13.6   Curreent State

The idea of Live Chat Support is very intuitive for any eCommerce website. It fits in the model of instantaneous access to information and is the next logical step in CSR customer interaction online. However, Instant Chat Services has not been adopted vary widely. This is mostly because the traditional phone CSR fulfills most of the functionality that Live Chat Support deems to provide. However Live Chat Support is best supported for sites which would require significant customer interaction such as customizable computers, sports equipment, home improvement items etc. where the customer would

> **Live Chat may be helpful if you rely on significant customer interaction.**

require an input from an expert on what item would best served for the customer's purpose.

eCommerce Best Practices

4.14 Blogs

  4.14.1   Introduction
  4.14.2   Fostering Customer Relationships
  4.14.3   Reasons to Blog
  4.14.4   Blog Netiquette
  4.14.5   Who Runs Your Blogs?
  4.14.6   Getting Started
  4.14.7   Monitoring Risk
  4.14.8   Feedback

4.14.1 Introduction

Information exchange is one of the greatest benefits the online world has been able to create. Instant communication regardless of location is revolutionary, and is changing the way to conduct business online. The popularity of blogs -- essentially online personal diaries open for outside viewing and commentary -- has expanded in the last few years as more and more people feel the urge to express their views with one another.

It was not long before companies found value in not only the information being gleaned from blogs, but value in adding a blog of their own into the blogosphere. The power of the blog is that it gives everyone from a voice and a platform to "say" what is on their mind. They can be CEOs or college students, but they both have equal means of expressing themselves through blogs.

Though blogs are typically created with the intentions to steer clear of commercialism, the open nature of them can prove to be advantageous to a company. Having direct discussions and conversations without the need of going through a middleman may help people better understand your company and get a full picture of your values.

> **Communicate *with* other bloggers not *at* them. Blogs are not advertising space, so don't treat them as such.**

4.14.2 Fostering Customer Relationships

Technological advancements are making communication a lot more accessible and convenient than ever before. It used to be that the only interaction between a customer and your company was through a cashier or a customer service

representative. Now, interaction can go a lot deeper because it delves into customers' personal time spent on contributing to blogs.

Because you are engaging with customers or potential customers when their guards are down, you are able to form a closer customer relationship that goes beyond supply and demand.

Creating a blog is a way to put a face to your company's name. It provides an opportunity to inject personality into your brand so that it is relatable on a human level. Understand that it is a way to directly communicate with your customers and not at them. Just like any worthwhile relationship, you need to invest time in building and nurturing it. If you do not have the time or the means of posting entries frequently—everyday or every other day—a blog would not be beneficial to you or your company. Customers are investing their own time into looking at your blog, so they take a vested interest into your thoughts, and if you do not give them regular updates you run the risk of alienating yourself from the community you started. Remember that bloggers are curious about you or your company and are actively seeking more information. They should not be treated like any old customer because their interest in you and your company is greater than that.

However, if you take the time to post entries and read people's comments or concerns you can build a community of loyal customers. You might not be directly selling your products or brand, but you are fostering a lasting relationship that highly reflects your brand.

But there are certain guidelines to follow when you post content to your blog because everything you say is subject to scrutiny.

4.14.3 Reasons to Blog

You will find that most retailers are absent from the blogosphere because they have trepidations in participating in a global online conversation. What they do not realize is that they are missing out on a key opportunity to talk to their public and listen to them directly.

Builds credibility. A well written and insightful blog informs readers about recent news, concerns, trends, etc. about your company or products. You can leverage this tool to mold readers to trust your thought leadership, thus producing more loyalty.

eCommerce Best Practices

Makes you Approachable. Shoppers will relate to your brand with more conviction once they understand your philosophy, values, and attitude is accessible and open-minded. Your blog should be more "real" than the corporate site.

Generates more links. Search engines love blogs because each user creates interdependency links within their posts. Blog posts are also syndicated to other technologies like RSS, and thereby increases listings.

4.14.4 Blog Netiquette

A big mistake when creating a blog would be to assume you can rant and rave about anything and everything. Though the blog could be run by one person within the company, remember that it is more a corporate blog that reflects your brand. But despite that there is etiquette, or netiquette rather, that is not taken lightly in the blogosphere.

A blog, corporate or non-corporate, is not a space for hard selling or advertising. Other bloggers could be put off by this because it is supposed to be an environment of sharing stories, thoughts, ideas, and concerns with one another. They do not want to be bombarded with newsletters or articles praising your company, or ads and promos about your company. They are visiting your blogs to read the content of your character.

The content is vital—it is the heart of your blog. Distinguish your voice, tone and point of view from all the other blogs out there. Language is key—you should be honest and open with your thoughts, however that does not mean you should not censor yourself, especially because your company is directly or indirectly involved. You do not want to have an "insert foot in mouth" moment if it can be avoided. Also, avoid formal and calculated verbiage that sounds like a lawyer drafted it up. Embrace spelling mistakes or harmless grammatical errors because it adds character and shows that you are not inhuman.

> **Do not avoid conflict, address it head on. Blogs are great platforms for clearing the air.**

Blogs can be a big asset to you in times of crisis. Should something occur to your company that makes significant waves in the media, a blog provides a platform for you to address it to the people that care most. When problems arise, do not ignore them as it might make other bloggers lose respect for you and your company. Instead, address it head on and be upfront with people about the situation. Provide explanations or offer apologies when appropriate, but also put a positive spin on it. Candor goes a long way, and people are more easily forgiving when they are told the truth.

4.14.5   Who Runs Your Blog?

You may be surprised to hear that many CEOs and executives run their own blog. Ted Leonsis of America Online, Marc Cuban of the Dallas Mavericks, and Bob Lutz of General Motors all have their own blogs. The question then arises as to why they would invest time into it? Simply put, it is value added to their brand. It is an inexpensive way to interact with customers or fans and it allows them to voice their opinions on things, and it also makes them more approachable, which makes their brands more relatable.

When executives run their own blogs, it shows other bloggers that these upper crust guys are investing time to get to know their audience. However, if you decide to hire a specific person or team to run and monitor your blog, make sure your selection will serve your best interests. Ideally, you want people that fully understand your brand and are passionate about seeing it succeed.

> **Choose reliable, passionate people to run your blogs so that they serve your best interests.**

If you run a company that sells cars or clothes, or makes furniture or Tupperware, whatever the case your blog will attract people with questions on their minds. If you cannot answer these questions, and it is important to answer every question, this is where hiring a dedicated staff that can answer those questions becomes important, especially if the blog has a large following.

If you come across other bloggers that love your brand, you might not need to go far to find the best people to represent your brand. If you have enough trust in those outside sources, and if they have established themselves as reliable ambassadors of your brand, then they might be willing to run your blog without cost or for free products or discounts.

4.14.6   Getting Started

Blogs can be so powerful that it adapts a culture of its own and can change the way a brand sees itself. Here are some options you have when starting a blog:

- Use a blogging software or service. Blogger.com, Livejournal.com, Typepad.com or Blogspot.com are all hosts of blog sites. Alternately,

eCommerce Best Practices

you can install blogging software like WordPress or Moveable Type on your server and link it to your retail site.

- Insert interesting links that are augmented with brief and clear commentary. Remember to update these frequently to keep users interested. You can also hire a writer or a skilled story teller to consult in how to better frame the content in a creative way.

- Provide multiple paths to your posts so that users can find old posts and categories without a navigation or search hiccup.

- Once the site is up, you should consider having an RSS feed, which is generally built into the blog services. RSS feeds are hosted on the blog's web server and are subscribed to searches, often without additional costs.

- Register with other blog communities to generate more traffic to yours through outside links — this also increases your Search Engine visibility.

- Engage in user participation and allow them to subscribe to updates with their email address or an RSS feed.

4.14.7   Monitoring Risk

Because blogs are so open in nature, people are going to criticize you and have negative things to say and, unfortunately, you cannot control that from happening. However, there are preventative and censoring methods of controlling what gets seen or filtered out. It is not recommended that you filter out every negative comment about you or your company because you have to be able to take the good with the bad, and if the only things that are displayed on your blog are glowing commentary then red flags are bound to go up surrounding your blog.

That said, if commentary is deemed too graphic, inappropriate, derogatory, insensitive or libelous then measures need to be taken in order to remedy the situation. You can either monitor messages internally through filtering software, or outsource the monitoring all together to a third party. Whichever way you decide, the filtering software works so that it scans incoming messages or posts before they can be viewed and the messages that are inappropriate can be disregarded. There is also software that searches for content from messages that have already been posted, and can delete them as necessary.

4.14.8 Feedback

One of the greatest benefits of blogs is that you can acquire previously unknown information or insight from your fellow bloggers. Knowledge sharing is extremely important because you can learn what is top of mind for customers and what should be avoided. What they tell you can boost your spot in the market.

Their concerns are your concerns. Listen to them because they can be your biggest cheerleader when it comes to word of mouth or word of mouse. You might be able to predict trends or prepare for a higher level of in stock products based on what other bloggers say.

In addition to user feedback, there is the opportunity to read the comments and posts to your competitors' blogs (and vice versa). You can keep tabs on how the users on their blogs interact differently from yours. If you find that their blog is generating more popularity, ask yourself what are they doing that you are not? And what can you do that challenges them, creatively or functionally?

Think of blogs as the new bulletin boards. They are intended to drum up interest about topics or items. Blogs create greater opportunities for direct communication with your customers. Their popularity is due to the disregard for corporate politics and a refreshing honesty that they do not typically hear from companies. Though this is not the place for hard-selling or advertising, it is a place to get to know your customers and relate to them in a more personal way. When people take the time to understand you that concern gets reflected on to your company. So long as you stay within the blog netiquette, you can use them to build meaningful relationships with customers and gain needed insight for your future.

eCommerce Best Practices

4.15 RSS / PODCASTS

4.15.1 Introduction
4.15.2 RSS vs. E-Mail
4.15.3 RSS Best Practices for Feed Producers
4.15.4 RSS Best Practices for Feed Consumers
4.15.5 Podcasting
4.15.6 RSS and eCommerce

4.15.1 Introduction

Really Simple Syndication, or RSS, has permeated into every part of the World Wide Web and, as its name suggests, its simplicity is its selling point. In this day and age of complex web application interfaces and even more complex web pages that can consist of a maze of hyper links and, not to mention a plethora of advertisements, it is easy to see how RSS has gained a foothold. RSS is a simple and effective way of disseminating information, which can be seen in news websites as almost every major news website offers users the choice to subscribe to their RSS feed. A popular example of this is the site Bloglines.com which offers RSS aggregation features and supports itself by injecting small text ads in between the RSS feeds. In eCommerce terms, RSS has become a valuable marketing tool as most websites who have weekly sales events offer up deals for RSS feeds that users can subscribe to.

4.15.2 RSS vs. Email

RSS has an edge over email since it is based on user selection and allows the user to choose the kind of content they will see. This allows for marketing mechanisms that bypass problems with spam filters that block emails. These days almost all browsers support some form of RSS aggregation and there are a variety of RSS feed aggregating desktop software including numerous email clients.

> **RSS is driven by user selection and gives them the power to choose the content they want to see.**

4.15.3 RSS Best Practices for Feed Producers

RSS 2.0 specification is the dominant method of RSS distribution via XML. However, it is plagued by many interoperability issues and the specification is considered ambiguous by most developers. ATOM 1.0 which draws from RSS 2.0 is the specification supported by the IETF and is considered the better specification.

- Use of a feed validation service early and often is highly recommended.
- Ensure unique IDs for articles.
- Support autodiscovery; do not use text/XML content type.
- Use atom:summary for summary data, atom:content for full content.
- Embed well formed XHTML.
- An XSL style sheet is the best method for formatting XML output.
- Using an RSS publishing website such as Feedburner is always a good idea.
- Encourage the practice of embedding license metadata in the feeds.
- Use ping services to notify third parties about feed updates.
- For decent interoperability with reasonable security, use HTTP basic authentication over SSL.
- For excellent interoperability with low security, use obscure feed URLs.

4.15.4 RSS Best Practices for Feed Consumers

Accept at least Atom 1.0 and RSS 2.0. Consider supporting other variants. Use a library or proxy to normalize all feed formats for you—the ROME library is good example for doing normalization in Java.

- Support for autodiscovery is very important.
- Check for license metadata and perform the needed processing to honor those licenses.
- Support HTTP conditional GET so that the RSS data will be refreshed only if the content has refreshed.

> The enclosures tag is an effective way to add rich digital media content to a feed.

4.15.5 Podcasting

Podcasting is merely using RSS's capabilities to provide audio and video content. It has taken off in the past few years, mainly because of the deluge of portable music players in the market. The key to providing an effective podcast service is to recognize the target audience and decide on the right audio

eCommerce Best Practices

format. Currently MP3 and MP4 are the most popular formats for audio and video, respectively.

The most important feature that RSS 2.0 provides as far as Podcasting goes is the enclosures tag, which is a way of adding rich digital media content to a feed. We should also keep in mind that Apple's iPod is the most prevalent portable music player in the market, thus coining the term "podcast," and adding information using the iTunes namespace is a good practice. The iTunes namespace provides a series of tags that are specifically designed to be compatible with Apple's line of iPod players.

RSS is a powerful and simple way of disseminating information in a non-email dependent manner. The increase in the number of RSS clients, including many popular browsers and email clients is only helping the spread of this technology. Podcasting is another RSS off-shoot which has become popular in the past few years. Because many websites such as Podzinger.com offer podcast searches based on the podcast's actual content, podcasts are likely to grow at a faster rate.

4.15.6   RSS and eCommerce

As of the writing of this book, there has been almost no strong tie-in between RSS and eCommerce. However there is an opportunity to take advantage of RSS to keep a customer involved in your eCommerce site. Listed below are some of the ways that RSS can be used to encourage customers back to your site.

Provide RSS for new products for each category: Customers are often interested in finding out the latest and the greatest your site would have to offer for various product categories. An example of this would be new DVD or Blue-Ray movie releases. By providing an RSS feed of the new releases page, the customer can quickly scan the feed as part of their daily routine and make a purchase if they so desire. Give users multiple levels of RSS feeds e.g. allow users to subscribe to new DVDs for Action, Comedy etc.

Provide RSS for new products for search queries: Some user interests might not cleanly fall under any particular category, such as "Kid friendly cups". Allow users to create an RSS feed for search results so they can keep up with new products within your site.

RSS feeds for tracking packages and returns: Allow users to subscribe to an RSS feed which is regularly updated with the status of their orders and returns. This would provide more information that an email informing them of their shipment, but would provide each location their order shipment is passing through. This would increase user interest and excitement in the products they have purchased and have the user more involved with your website.

4.16 Portals

4.16.1 Introduction
4.16.2 Modernize Your Business
4.16.3 Flavors of Portals
4.16.4 Benefits of Deploying Portals
4.16.5 Portals and Their Impact on an Organization
4.16.6 Layout of Typical Portal Implementation

4.16.1 Introduction

Essentially, portals are next-generation desktops. A portal offers a cohesive platform to access the applications and content in secured collaborative manner. A portal is a framework supporting a broad assortment of applications from intranet applications, extranet applications, database application, and secured and collaborative applications. Portals are designed to meet the goals of organizations of any size — small, medium to large organization. Portals can offer most scalable, secure and robust communications within an organization.

A portal can offer dependable, integrated user experiences that provide a single view of components within the context. This is a very powerful concept that enables the consistent and integrated views of multiple components and services for the portal end user. Portals give the flexibility to end users to personalize or customize the portal based workspaces.

> **Portals can enable consistent and integrated views of multiple components and services.**

The portal application provides good front end support in Service Oriented Architectures (SOA). The open framework portal offers a basic building block called Portlets. Portlets allows the development team to focus on the unique aspects of their application, while the middleware handles the application life cycle, transactions and integration of components.

Portal framework is flexible enough to integrate the applications that were originally designed separately. With portal technologies the portal administrator becomes the application integrator, without much programming, by defining new portal pages, and adding the portlets to those pages. Portal end users can become their own application assemblers by personalizing and customizing the portlets, however, the portal administrator can control the end user access to customize the portlets. Some of the key benefits for using portals for your organization are that it can increase employee productivity, build customer loyalty, introduce collaboration capabilities between disparate processes, and increase integration with back-end systems.

4.16.2   Modernize Your Business

Getting the right information in a timely manner is a constant organizational problem. It involves individual effort and causes a lot of stress. Finding the right people who can help requires an extensive amount of work. The intranet offers a wealth of information, but it takes effort to mine the required data. Getting the right information to right people at right time is the purpose of portal application. Transform your business into a successful, collaborative business by implementing the Portal application and Personalization.

Portals are software that can customize information (content) and transactions to end users. Unlike an intranet or extranet application that only helps organizations distribute information on the web, a portal application also lets users integrate and run applications such as tracking a shipment or pending invoices in a supplier's ERP system.

> **Customize your portals to fit your business purpose— there isn't just one solution.**

Most portals are focused on employees and partners; collaboration is the main theme behind these. Portal solutions can be designed to fit the needs of particular target audience or business segments. There are business-to-business (B2B), business-to-employee (B2E), business-to-consumer (B2C) and collaboration portals.

Business-to-Business (B2B) Portals
A B2B portal solution facilitates communication between company employees, business processes and suppliers. This type of portal is also known as extranet portals. It is designed to result in a shorter business cycle, so that suppliers can check their order-to-delivery cycles, cleared payments and employees can track orders, shipments, etc. This type of portal also results in an increase in productivity and a reduction in bottle necks as the information is readily available.

Business-to-Employee (B2E) Portals
A B2E portal solution is designed to increase productivity and reduce costs by providing a single interface for the necessary applications, processes and information. This type of portal is also historically known as an intranet portal. The following are some of the functions of B2E portal HR, payroll, vacation balance/request, etc. This type of portal is designed to integrate existing applications, increase the collaboration among the departments or the people within the organization and reduce complexity. The benefit of employee intranet is faster access to people, processes, information and improve communications.

Business-to-Consumer (B2C) Portals
A B2C portal solution is designed to support better customer service and increase revenue. This type of portal helps businesses be more responsive to customers and provide the right level of information and up-to-date information while gaining the customers confidence and retaining them as loyal customer to the business

Collaboration Portals
The collaboration portals are designed to facilitate teamwork, communication and human interaction. The goal of this type of portal is drive productivity to a higher level among geographically dispersed teams to work as though they were all in the same location. This type of portal supports content management services (CMS), extracts and organizes related information along with collaborative services like chat, email, shared calendars and user defined communities. Collaborative portal installations are typically bounds within the organization.

4.16.3   Benefits of Deploying Portals

Businesses choosing one of the portal projects can realize concrete business goals and technical benefits.

> **Increased Productivity** – This is because access to more relevant information is available in a single access point to applications and collaboration tools.
> **Better Integration** – Enables faster integration with backend applications and business processes in a single interface.
> **Single Sign On** – Fewer passwords to remember and administer gives better user experience.

Secured Access – Portals have security and access control as most organizations need to control the information based on the user is entitle to see.
Personalization – Delivers up-to-date personalized content and processes for your target audience.
Collaboration – Powerful collaboration capabilities like instant messaging, discussion board, alerts, shared calendars, etc.
Training Cost – Reduced training cost resulting from common presentation and a consistent user interface.
Cost Benefits – Close relationships with customers or partners results in workforce productivity, innovation and reduced cycle times. Additional cost savings through website consolidation.
Rapid Time-to-Market – Since a large amount of an organization's intellectual property is re-used in different contexts, the time it takes for marketing efforts is reduced.

4.16.4 Portals and Their Impact on an Organization

Portal solution provides an integrated service across your organization's functions and business units. Therefore, there is no sole owner for all business applications—the portal solution team has shared ownership and accountability for strategy, development and operations. Alongside the technology, content, and services, successful portal implementation includes governance. Governance is needed not only during the initial phase of the project but also during the ongoing expansion.

> **Governing the design portal is vital for staying true to the business requirements and solutions.**

Therefore, it is important to establish a committee who will govern the design portal. Design includes, but is not limited to, security, navigation and system integration. The governance committee will work alongside architects to prioritize and validate business requirements and to review feasible solutions. The governance committee should carefully define which content needs to be centrally controlled and which content can be safely handed off. There are a number of things that should be defined and managed centrally; here are some of the examples:

Security and Authentication
Since most organizations need control over the information and decisions made at various levels of management, it is essential that portal solutions reflect this.

Technologies like LDAP and single sign-on are mostly managed centrally within the organization. Portal framework and its application could use this method rather than implement its own user directory. Most of the enterprise information portal (EIP) solution available in the market can leverage existing directory services in the organization.

Enterprise Branding (Look-and-Feel, Navigation and Design)
Often a great deal of time and effort is spent during "local" portal initiatives to create portal branding, a better look-and-feel, and efficient user navigation. While an organization should encourage creativity within a community's portal, a branding style guide and templates should be provided based upon their previous portal experience.

Community Organization and Definition
Community by definition means a group of users with a common interest, e.g. the Human Resource department or Manufacturing department. Initially a centralized group within an organization would create the definition of the core set of portal applications and communities, which most likely correlate with the main functional areas of an organization. There may be layers of communities that lie below this top-level, but the initial structure needs to be agreed upon from somewhere high up.

Metrics/Scorecard
Every portal should have a defined set of metrics by which the success is measured. The measurement of such metrics should be implanted into the structure of portals, and the reporting and analysis of these metrics should ideally be performed centrally.

Content Development and Management
Content developed for EIP portals may need editorial and approval procedures per the content defined at a corporate level. The content will be used within the community portal. Portals can leverage the workflow capabilities of the content management system (CMS) tool that your organization may have rather than creating the portal's own workflow.

4.16.5   Layout of Typical Portal Implementation

Portals help people in the way they work, communicate and collaborate. The centralized controls should allow community leaders and users enough autonomy to assemble the portal at the community level.

eCommerce Best Practices

Pages and Layouts
The design and layout of the portal page should be delegated to the central governance or community owner, as they will most likely investigate the creation of the portal in the first place. Hence, they can have some notion of what needs to be achieved. For example, community owner can make a decision about whether or not to allow end users to customize their versions of each page.

Templates
In order to easily create new portals, there needs to be a template from which they can be formed. From the template, portals can be quickly spawned as necessary. The template definition contains the pages and portlets that will be in the new portal solution.

Themes / Styles
Themes provide the navigation, appearance and layout of the portal, including colors, fonts and images outside of the portal content area (header with logo or banners etc).

Portlet Usage
By definition, a portlet means a reusable standalone J2EE application that can be placed on the portal page to perform a specific function. As part of creating a base portal, a central governance committee may allocate a set of portlets that can be used within the community pages, but it is the community administrator or leader decides which portlets will be used and how they will be configured.

Community Membership
Each community could have an administrator role called community administrator, who can administer and manage the membership. As mentioned earlier, portals could potentially use the corporate user directory. The community administrator should be able to manage users, role and member organizations.

Entitlements
The community administrator decides who should be member of a community and who has the privilege to access the content within their community. Portals can delegate fine-grained permission to community members. Even if the portal structure and layouts look similar, every community will have its own user roles, responsibilities and permission levels. For example, not everyone has access to all portlets. By being able to define permissions at this level, portals can serve targeted content to its users in a single interface.

Delegated Administration

To successfully manage portal, community members need to have sufficient roles and responsibilities delegated to them by the central governance. The community leader should be responsible for developing their community, which includes customization to the portal, manage members and roles, and granting permissions.

4.16.6 Shared Portlets and Community Portlets

An organization could potentially have more than one set of users through a departmental or skill set segmentation. In order to keep a good level of consistent information across these disparate sets of communities, a community template definition should give careful consideration to which portlets can be "shared" across communities and which portlets cannot be. If there is common information, content can be encapsulated in a shared portal which could be shared across many different communities. The fact that portlets may be shared does not negate the possibility of personalizing the content that it displays. But rather, the Personalization rule can be defined such that a shared portlet can show different content to different users based on their personal preferences. For example, one community member might like to see 10 results in page, other one might prefer to see 20 results. The combination of personalization rules and portals gives the organization considerable control over the delivery of its portal content.

> **Portlets can be personalized to show different content to different users.**

Portal solution can determine which portlets (elements and functions) are specific to a community and which portlets can be shared by other communities, or can target the content for a specific target audience by a personalization feature. This will allow the portal owner to focus on the content that is relevant to their community. Also, portals offer a way to balance centralized control and local administration of portal applications. Many business models have requirements to empower a community of users to be autonomous while ensuring the functions within a pre-defined framework. Portal solutions work well within an organization or extended out to business partners, affiliates or other trusted parties because they offer organizations an extremely powerful tool to empower and distribute the management of their portal for areas such as multi-tiered distribution models, or partner relationship management. All of these solutions result in cost effective ways of managing large numbers of portal users within a single infrastructure while reducing the cost of ownership by distributing portal management.

eCommerce Best Practices

V
how-to

ecommerce
bestpractices

Part V: How-To

eCommerce Best Practices

HOW-TO

It is one thing to decide to market your organization's goods and services online, but actually implementing and managing it to be a revenue generating vehicle can be an overwhelming and costly endeavor. In fact, most organizations fail to recognize that implementing new technologies requires more than just an IT-funded initiative, but rather a demand-side sponsorship that is actively involved in the requirements defining period. It is this salesforce and customer involvement that is crucial in accurately defining the business needs that will ultimately evolve into the selection of the proper toolset, skilled resources and environment infrastructure required to make the eCommerce initiative a success.

From a marketing perspective, the degree to which an organization chooses to integrate content management, manage distribution channels and product costing directly affects the customer reach and future scalability of their eCommerce potential. Depending upon market conditions and competition landscapes, integrating all "must-have" functionality will not necessarily provide an immediate positive impact to the bottom line, but may rather, result in a catastrophic cost bleeding if the correct corporate-wide support is not in place. Again, market externalities aside, it is in the organization's best interest to identify all "best-fit" features for initial rollout and group additional needs for future phases. This provides the organization ample time to manage internal change and marginalize customer sticker shock.

As part of the How-To portion of Best Practices, the following items were identified as most applicable and routine during an eCommerce initiative rollout:

- RFP Considerations
- Implementation Considerations
- Importance of Operations/Maintenance
- Data Migration Properties
- Content Management Features
- Reducing Fulfillment Discrepancies
- Holiday Season Tactics
- Potential of Multi-Channels
- ROI and Cost Analysis
- Rough Order Of Magnitude (ROM) Estimation

eCommerce Best Practices

5.1 RFP

   5.1.1   Introduction
   5.1.2   Checklist in the RFP Process
   5.1.3   Flaws in the RFP Process
   5.1.4   Sample RFP
   5.1.5   What Limits Options in RFP
   5.1.6   RFI as Solution

5.1.1 Introduction

Issuing a request for proposal – or RFP – has become a popular practice in the retail and direct marketing industries as a mechanism for maximizing initial project cost savings by creating a vendor competitive landscape. When developing an RFP, organizations should be aware that they are defining a relationship that could span many years and possibly touch various departments outside of the initial scope. Ideally, RFPs compel the vendor to offer the lowest-cost and best-match to an organizations business requirements. They also help organizations uncover any operational bottlenecks or challenges outside of the project's impact zone that will aid in the vendor's solution approach.

5.1.2 Checklist in the RFP Process

The entire creation process of an RFP is a fairly standardized procedure that ensures a level playing field amongst all parties involved. While crafting an RFP, consider the questions to ask all vendors that will help you filter out the most qualified party. Successful RFPs are dependent on buyer knowledge, usually based on direct experience. Some of the points one needs to consider before creating an RFP are:

Know if an RFP is Necessary

The ability to deliver a solution by a required go-live date can be put in jeopardy by the process of documenting, debating and selecting a vendor. If time-to-market is a factor, the organization can conduct brief requirements gathering and select best-of-breed vendors based on market competition. Although this is not best practice as it can result in higher implementation costs and a product that is more advanced than required,

eCommerce Best Practices

a scaled down version of the solution can be a valuable quick-win for the business.

Establish Vendor Communication

Open discussions between organization and vendor can significantly reduce the time and effort involved in the review process. Points to consider during these discussions meeting may involve pricing considerations, must-have functionality, and knowledge of your competitive landscape. Cultural match between vendor and organization is a critical component in delivering a solution; however, it should not be a show-stopping decision factor during the review process.

Clearly Defined Budget

Before entering into the negotiation phase of an RFP, it is critical that the organization has their internal floor and ceiling price points clearly identified. These are the two most valuable components of information during the negotiation phase of the RFP. It is a fine balance between sharing your spending criteria with your vendor to ensure your wants and needs are met and risking overpaying for services rendered.

Define Needs; Not Solutions

The RFP details only have to cover your current and future business needs. It shouldn't propose or suggest solutions that the organization may or may not favor. The RFP should be crafted to for the vendor to address how they would creatively solve your business challenges. An effectively written RFP clearly articulate the organization's objective, strategy, and desired goals that need and even want to be reached.

> RFPs should define the need and not the solution when addressing vendors. Requirements should not be tightly defined to prevent vendors from being innovative.

Best-Fit Vendors

The RFP should evaluate the type of vendor desired based on how they most match your business's "must-have" requirements. If you find that it heavily relies on IT resources and requirements, you should consider focusing on selecting vendors with IT management expertise. The same holds if the RFP is for systems integrators, online marketing, etc. You want to make sure that the vendors you contact with an RFP are suitable and relevant to your needs.

Be Prepared for Selecting Multiple Vendors

Most of the vendors today specialize in one aspect of the greater implementation of the eCommerce process. Because of this, there very well may be more than one vendor helping you achieve your online objectives. Best practices suggest that the vendors you shortlist and end up choosing are held responsible only for the tasks that are specified in completing. From a contractual standpoint, it makes sense to request that multiple vendor's team together as one agent of delivery. Aside from the obvious legal benefits, this provides the most protection for the organization when it comes to setting and exceeding delivery expectations.

Judgment Process

Once RFPs are submitted and final vendor responses are captured, in-house judgments should be based on the vendors' overall approach to the problem, creativity or technical savvy in addressing the needs as mentioned and your sense of their reliability, responsiveness, skills and, of course, price in relations to the tasks at hand.

5.1.3 Flaws in the RFP Process

With the checklist in mind, clients can be reasonably sure of a painless writing process. But, there are also some serious flaws which need to be considered:

- Submitting an RFP to vendors without informally communicating with vendor representatives first is not advisable. It is important for retailers to check before putting both you and a potential vendor through the rigors of producing and judging an RFP.
- Making judgments based on price alone will not help you attain the right vendor. A vendor's understanding of your business, your requirements, and their creative and rational approaches are some aspects which need to be weighed out against the price bid.
- Be careful of inviting vendors who are not serious about addressing your business needs in order to increase competitive price bids. This can quickly result in wasted resources and time during the selection process.
- Requirements should allow room for creativity, but not at the risk of creating organization chaos during implementation. A tightly defined prohibits the vendor from presenting an innovative, out-of-the-box

eCommerce Best Practices

solution and also emphasizing their strengths and value to the marketplace.

5.1.4 Sample RFP

RFPs can be written for a broad range of requirements from all the departments of the retailer. But, that doesn't mean that every RFP is different from the other. To a certain extent the content and certain requirements would differ but all the RFPs include the same pre-defined sections like purpose, overview, RFP response, business requirements, technical requirements, support requirements, pricing and costs and finally summary.

> RFPs should always detail: Purpose, Overview, Response, Business requirements, Technical requirements, Support requirements, Pricing and Costs, and Summary.

Purpose

An executive brief or quick read that summarizes the entire RFP in a few sentences. The reader should be able to able to discern the goal of the project at a high level. This section should carry a short description about the purpose of the RFP which is simple and straightforward in its objective. Retailers should be aware that they are requesting a proposal from multiple vendors for a possible service relationship.

Overview

Also a brief and simple section this should mainly cater to inform the vendors who you are as a company, your business, and how you want the vendor to handle the functions that they are outsourcing currently.

RFP Response

This section requires being more substantial in content. The retailer should request a response for addressing the vendor's product history, what their implementation plans are, process for customization and training and finally customer references.

Business Requirements

This includes detailed information about how the proposed solution would operate within your organization's environment. Care should be taken to ensure that the correct business requirement information is targeted and specified with respect to all the needs you want met.

Technical Requirements

This includes any technical details that are pre-existing in your organization that will determine your end ability to reach your end goal. All question and answer slots should be made here so that the vendor can complete answers or, if needed, ask you for any clarifications. System details, management and requirement technicalities are some specifications that can be added to the list.

Support Requirements

This involves asking the vendor what specific customer support would be provided for the production environment: email, telephone, etc. Address details such as 24x7 support for all channels to help minimize the response time to customer queries, requests, back orders, etc.

Pricing and Deadlines

Be straightforward about your pricing format so as not to confuse the vendors or mislead them in any way. Identify startup costs, additional support costs; customization costs and, also in particular, any hidden costs. Deadlines for submission are equally important so that there is no delay in the process for finalization. All deadlines should be specified and valid points of contact should be identified from your side. Deadlines are a big factor in determining whether a vendor can actually deliver tasks on schedule.

5.1.5   Limitations to RFPs

The RFP process limits open dialogue between the vendor and client. Usually communication with the vendors is formal and restrictive—often limited to the answers for the set number of questions asked in the RFP. However, best practices suggest that vendors are given a chance to ask questions that can clarify their doubts, but also in some cases add more value to points that were originally asked in the RFP. This generally occurs during an open forum period where all vendors are invited to attend, and given an opportunity to ask probing questions.

RFPs limit the dialogue for an open Q&A. However, there are cases where vendors can ask questions only if they are willing to share their questions and answers with the rest of the competitors submitting proposals. Here, some vendors with high competitive

> **RFPs reduce the dialogue for open discussions to level the playing field, but should questions arise, make sure to share them with all parties.**

distinctions are reluctant to disclose the information to their rivals. Measures are taken to ensure a level playing field for all parties involved. This makes it easier for you to make the right decision once all the vendors' responses are evaluated.

5.1.6 RFI as a Solution

Request for information or RFI is an alternative solution which can be used for establishing an initial point of contact with a vendor before the RFP is sent. Initiating a dialogue with vendor through an RFI has an advantage. Information gathered here lays the foundation for a more clearly defined RFP in the future.

A well-written RFP that clearly states your eCommerce intentions and goals will make the vendor selection process more productive and less taxing on internal resources than an RFP written in general terms. In doing so, this reduces confusion on the vendors' behalf and gives you a clearer reference point to assess the vendors' answers. Once the vendors are short-listed, it is always in best practice to visit them face-to-face and observe their work environment.

eCommerce Best Practices

5.2 Implementation

 5.2.1 Introduction
 5.2.2 Design Patterns
 5.2.3 Reusing Code
 5.2.4 Automation Tools

5.2.1 Introduction

The implementation phase is the very heart of any eCommerce project because it is the process that turns requirements into an actual working solution. Unfortunately, without the correct resource management and cross-departmental support this process can be quite chaotic, and may produce unreliable results. An eCommerce project is not necessarily a one shot, fail or succeed, opportunity. Due to its direct customer exposure to an organization's product line and logistical operations, the failure of an acutely designed implementation can cause severe brand damage. Thus, it is important for the project management team to follow a process that is known to maximize the chance of success by minimizing cost and time and by maximizing quality. This process is known as an implementation methodology and the rigor with which it is implemented is vital to the success of full life-cycle projects.

In 1676 Isaac Newton explained to a contemporary that one of his discoveries was only made "by standing on the shoulders of giants." This notion of building upon the great works of others is the central principle behind good implementation methodology and – not coincidently – it is the cornerstone of modern software design. Almost any system currently in deployment relies, at least in part, on previously existing software. And, reinventing the wheel is costly, time consuming and will likely result in a buggy wheel. This principle can be applied across many domains: in software design it is called design patterns, in software development it is called code reuse, and in the development environment it is a set of automation tools.

> **A design pattern is the industry best practice for solving a problem; therefore, it is the most commonly implemented solution.**

5.2.2 Design Patterns

While a discussion of design patterns may be more appropriate during the analysis phase, the implementation team must recognize, apply, and employ

certain patterns even when they are not specifically called for in the technical design document. A design pattern, then, is a "pattern [that] describes a problem which occurs over and over again in our environment, and then describes the core of the solution to that problem" [Alexander et al. 1977]. In other words, it is the industry is best practice for solving a problem, and, therefore it is one of the most commonly implemented solutions. In the most common situations, it must minimize cost and time and maximize quality.

Using a design pattern allows a complex construct to be explained unambiguously and implemented with a much shorter ramp-up period for the implementation team. This ensures a higher quality product at a lower cost and over a shorter time period than would otherwise be possible. The same benefits can be realized when design patterns are built in to an API specification.

5.2.3 Reusing Code

Code reuse is the bread and butter of modern software engineering; it is the practice of reusing existing code in the solution to a new problem. This includes using existing libraries, repurposing an existing project, or interfacing with an external module or service. Reusing code speeds up development because any code that exists has already been through the entire software development cycle at least once. Therefore, it required quality checks and testing time can be decreased and it has few enough bugs to meet the quality control standards of the organization that developed it. In essence, a project that reuses code can be thought of as having more developers assigned to it than a project that does not.

Besides the obvious time-savings that this provides, there can be other advantages depending upon the source of the code that is being reused. The Java software development kit, for example, includes a set of classes called Collections Framework that provides commonly used objects such as lists, sets, arrays, and maps. These classes have been extensively quality controlled in controlled environments as well as being used in almost every application that has ever been built with Java.

> **By relying upon Collection Framework, developers don't have to worry about reinventing the wheel, and can spend time working on the parts of the project they have the ability to optimize.**

By relying upon the Collection Framework, the developer is freed from worrying about how to write an efficient, bug free linked list and that extra time can be spent working on the parts of the project for which they are uniquely qualified.

eCommerce Best Practices

Thus, reusing code ensures that a product of similar quality can be delivered at a lower cost and in a shorter time than would otherwise be possible.

5.2.4 Automation Tools

Finally, the set of automation tools available to a developer can have a significant time impact on the implementation team's overall productivity. These tools can help debug, benchmark, deploy, or even generate code. Any tool that automates a common process for the developer can represent a gain in efficiency. Different tools have different benefits for the developer but the most common development tools are the Integrated Development Environment (IDE), the source control, the regression testing framework, and the build/deployment engine. There are dozens of IDEs available to developers and each IDE has a plethora of features that are supposed to aid the developer in one way or another, but the most common features are: auto-completion, code formatting/highlighting, and a debugger.

Auto-completion is a tool that offers suggestions to the developer about what methods or classes are available rather than forcing them to go back to the documentation. Code formatting/highlighting is a tool that helps reveal the structure of the code and improves its readability. The debugger is a tool that offers detailed "step-through" control over a running program so that the developer can ensure that the state of a program is valid as it executes.

Each of these tools saves time by making it unnecessary for the developer to leave the development environment to read documentation or to troubleshoot their program. Meanwhile, the source control module provides a versioned backup of the project's code and can help resolve conflicts between developers working on the same file. Source control is important because it facilitates a team of developers working in parallel and it provides a mechanism for disaster recovery. Without a source control module, a team of developers would have to synchronize their efforts manually to ensure that they are always using the most up-to-date code that is available and to prevent one developer from overwriting the changes of another. Moreover, if a catastrophic bug was introduced into the project there would be no way to roll back to an earlier version without the bug because there would be no way to ensure that anyone had the earlier version or even to recognize it if they did.

The regression testing framework and the build/deployment engine ensure that the code is always tested and deployed using the same procedure. Consistency in testing and deployment is important in recognizing defects and it is important to correct them in a timely manner. Each of these tools is designed to reduce the amount of time spent by the developer doing anything that is not development. They ensure that a product of similar quality can be delivered at a similar cost in a shorter time than would otherwise be possible.

Implementation is a critical phase of an eCommerce project because it is the backbone of a life-cycle process that turns the requirements into a live solution. Project teams thereby have to follow processes that maximize the quality and success of the implementation with minimum risks. Having clearly defined roles and tasks for project team members is critical in ensuring a more streamlined, less disordered implementation process.

eCommerce Best Practices

5.3 Operations / Maintenance

- 5.3.1 Introduction
- 5.3.2 Maintenance
- 5.3.3 Operations in Returns
- 5.3.4 Operations in Accounting
- 5.3.5 Operations in Marketing

5.3.1 Introduction

Operations management is an umbrella term for a business or organization's internal procedures for handling its workflow. An eCommerce site has multiple operations running together at any given time. Every operation is a piece of the puzzle, and an eCommerce site would not be complete if one was missing or working improperly. Depending on how customized your system was developed, your operations' processes could be very complex or very simple..

There is no set operations rule book that every business can follow, but because in eCommerce there are so many different types of operations working in tandem there needs to be a high level of organizational planning and execututive support in order for it to be successful. Ecommerce systems require maintenance in areas that help generate sales and customer consistency. You will find that many of the same operations that happen on an offline level should mirror the operations on an online level.

In the retail industry, eCommerce operations management can consist of supply chain management, inventory management, online transactions management, electronic funds transfers, UI, catalog management, customer service and call center, quality control, analytics, etc. Many merchants and retailers new to the eCommerce world are overwhelmed with the amount of coordination and work it takes to build a strong and successful website.

Housekeeping, updating, and upgrading alone can take an extensive amount of work depending on your scope and scale. It is important to keep your eCommerce expectations reasonable in regards to how complex your site needs to be built and how much you can grow as a business.

Catalog maintenance is a major challenge as it pertains to product availability. Because an eCommerce site needs to display images of products with a considerable amount of richness compared to a paper catalog, the amount of time and work involved in creating those images increase as well. It is not as simple as taking a photograph of a product and posting it to a page because if your online experience is as flat as a paper catalog than you are clearly not utilizing your eCommerce site's abilities to your advantage. Use a graphics application like Adobe Photoshop to clean up product snapshots and make the products more desirable — not to say to completely re-touch the image to the point where it does not even resemble the original.

Catalog management deals with the past, present, and future. It details the new products you will want to sell, the products you are currently selling, and the products you do not want to sell anymore. This operation should be fairly constant and ongoing for your eCommerce site. A product's life-cycle is shortening as more and more products get pushed, particularly in technology, and as the life-cycle decreases the rate of catalog change increases.

> **Carefully plan and time updates to current products, as well as the products for future seasons. This includes items to go on-sale, new product images and descriptions, etc. These can be time-consuming to create.**

When you update your product catalog, such as by adding products, shifting products to the sale section, or deleting products, make sure you have carefully planned and designated the products to go in what category before execution. Many commerce systems will let you upload product images and descriptions well before deployment, allowing you to store the completed products and campaigns for next season while working on the images and collections for the season after. This gives you more time to map out future product campaigns between the Product Development, Marketing, and IT departments.

Allow time for customer service representatives to learn what was put on sale and what was just added to the catalog as well. It is important to give them prep time and early word that changes to the product catalog will be made.

Corporate marketing campaigns like pricing strategies, assortment planning, and promotions are additional features that get included with general retail management issues like staff management, accounting, commissions and promotions management functions. In the eCommerce retail industry you should have various departments included within the operation and maintenance list — marketing, selling, catalogs, shipping, tracking and packing, returns, and multichannel.

Many direct commerce order management systems provide the retailers with a fully integrated eCommerce module or have an array of application programming interfaces (APIs) to web integration. But these same systems often do not provide more function to retail operations apart from the cash-drawer modules for warehouse stores.

Some retailers have a multiple point-of-sale (POS) cash register management system in multiple store locations that manage and replenish its own inventory either from a consolidated inventory serving all channels or from a dedicated inventory. Handheld devices here need to provide inventory tracking and counting functions as well

5.3.2   Maintenance

In most retail back end operations, maintenance plays a key role in ensuring the operations run smoothly, customers are supported and sales targets are achieved. Most retailers, particularly the small ones, lack the resources to employ the manpower and technology required to run the necessary maintenance. The problem is not that they are smaller organizations, but that they have other issues like scalability and sustaining the growth based on the technology provided.

Maintenance within the retail distribution center is more complex, with applications that include enterprise resource planning (ERP) systems, warehouse management and control systems, labor management systems, and transportation systems to name a few. Infrastructure like hardware, network, routers, and switches are more add-ons. The ability to connect operational output, like the warehouse control management system connecting to the warehouse management network, ERP systems provide management the tools to effectively plan the business.

The bottom line is that in order to effectively integrate systems and employ data sharing, it requires routine maintenance checks on a timely and periodic basis. If left unregulated, a seemingly small discrepancy in any one of these areas could reflect on the other, which may potentially snowball into a problem with great significance.

Software updates and releases are also a part of the maintenance process. Retailers should evaluate plans for systems maintenance prior to vendor selection. Different levels like operating system levels with security updates are crucial to the security aspect. In an application level, upgrade maintenance

adds features to improve performance and business; at network levels, spam upgrade maintenance ensures uninterrupted connectivity to the outside world.

In order to maintain the system's longevity and health, as the applications may be running independently on individual systems or in a shared environment, it requires purging and archiving routines to run automatically. This ensures that retailers keep only the mandatory amount of data in production environments, thereby increasing available disk space and system performance.

> **Make sure you archive and purge unnecessary data and run routine check-ups on your system.**

5.3.3   Operations in Returns

In eCommerce, or even brick and mortar commerce for that matter, product claims and returns are inevitable and unavoidable. The art is in how the organization is structured to manage them. An organization has the power to structure their return policy and has control over how smoothly their return process operates. Even the most relatively simple change in a return protocol can make a significant difference in customer satisfaction levels and bottom line improvements.

Retailers sometimes take on vendors who provide the return processing service for them. They pick up packages of the retailer from freight forwarding organizations and take them to the retailer's nearest processing center. Here these packages are scanned and reports are sent to the catalogers regarding the numbers and when to expect them at the distribution center.

> **Return authorization forms that give catalogers a heads-up on all the merchandise coming back to the distribution center helps streamline return process operations.**

Some retailers also invest in return logistics software where prepaid return labels are printed at the bottom of a standard shipping invoice. Generally, this software is versatile to include flexibility in the type of return items like segmenting items weighing less than a certain weight to be shipped via a selected shipping agent.

Customers also get return updates similar to the email alerts a shopper receives when placing an order. Step process in the emails include alerting customers that their return order has been entered into the retailer's system, their package has arrived at the distribution center, and credit is being returned to their account or an exchange voucher is on the way.

5.3.4 Operations in Accounting

Internet sales tax is often popularly misunderstood as it can cause confusion for both retailers and consumers. For example, when most retailers first start their eCommerce endeavors, they do not realize that they may very well be responsible for paying the sales tax on products sold. Customers, for that matter, may also need to pay the tax on orders (known as "use tax"). The complication of Internet tax is that there is thousands of varying taxable jurisdictions to keep track of and process for each transaction around the country, much less the world.

Generally speaking, if a person from one state makes a purchase from a supplier in another state, that purchaser should not have to pay tax on that order. However, if a person from one state makes a purchase from a supplier with locations or branches in the same state of the purchaser, then that person should have to pay tax on that order. Often, the idea of tax-free shopping sounds too good to be true and in fact it is. Whether it is from the retailer or consumer, tax is still technically owed to the state even if tax is not included in the order—regulating this, however, is another story.

Failing to pay these taxes can result in various penalties or fines. However, because the current Internet tax laws were created when online shopping was not as popular as it is now, the actual amount of taxes due used to be relatively small and enforcement of penalties were not very strict. Now that eCommerce has grown exponentially since its creation, the government is now looking for ways to recover this lost revenue. It is estimated that the amount of taxes owed in 2005 for Internet transactions alone was around $20 billion.

More states are now addressing this issue by requiring people to file their online transactions on their tax return. Because state revenue is so reliant upon their sales tax, approximately a third of its funding, states cannot afford to ignore the matter any longer. Although a streamlined method of implementing tax on online purchases is not strictly enforced, it is not something you or your customers can avoid. Make sure your operations can identify which users are taxed, and can report the taxes that they are owed.

5.3.5 Operations in Marketing

In eCommerce, an organization is not simply fulfilling customer orders — they are also fulfilling company marketing plans that will eventually be communicated to the customer. Open communication is critical to building a successful eCommerce operation, and this includes the Marketing team. The last thing you want to do is promise something you cannot deliver, and in order to avoid such problems the marketing team has to discuss and clearly define the operational goals for sales growth and quota fulfilled before launching a marketing initiative. Marketers need to know what new item to promote before they think of ways to promote it, and the way they promote it effects how they will forecast the demand.

New product catalogs require formal approval from business and systems operations before promotions are released — this is not only to find errors, but to learn about it ahead of time to prepare for customer expectations. Standard operating procedures help create a structured relationship between marketing and other operations, and serves as guidelines when developing a promotional plan.

Operations and Maintenance are part of building and sustaining a successful eCommerce site. Catalog management, inventory tracking and supply chain management, transactions, call center and customer service control are all involved in this process. With the above operations running in tandem, maintenance is equally important so that everything runs smoothly. The level of complexity within a retail distribution center depends on whether it includes an ERP system, warehouse management, and an IT infrastructure with hardware and software to run the logistics and delivery management. Special attention and focus needs to be implemented during seasonal times with regards to marketing, customer service, returns and accounting. Only make promises you can deliver, so minimize the mistakes by establishing a sound operations and management system.

5.4 Data Migration

5.4.1 Introduction
5.4.2 Tools for Data Migration
5.4.3 Issues with Data Migration
5.4.4 Best Practices for Effective Data Migration
5.4.5 Planning

5.4.1 Introduction

Data Migration is the set of activities that safely transfers data from one or more legacy system or application to a new system or application without harming data integrity. The main aim of the data migration process is to preserve core business knowledge to make it accessible from the new application. Retailers should know that the process of data migration is a long and complex task that involves changing the entire IT infrastructure and should only be executed in controlled environments outside of production. It is advisable to avoid data migration during the holiday season as there is no guarantee or best practice that ascertains a complete hassle-free process, so leave enough time for the system to adjust and run smoothly.

There have been cases where marketing and logistical departments were not aware of the data migration, which resulted in misplaced orders and conflicting data extraction. Make aware when certain data migrations will take place to avoid extra complications. To approach how a standard data migration takes place, you need to know the different stages it will pass through.

Data migration involves the following major stages:

- Planning and scoping of the migration process.
- Extracting data from the legacy system.
- Cleansing the data.
- Transform data to Load the data into the new system.
- Validate the migrated data.
- Verify the accuracy of the migrated data.

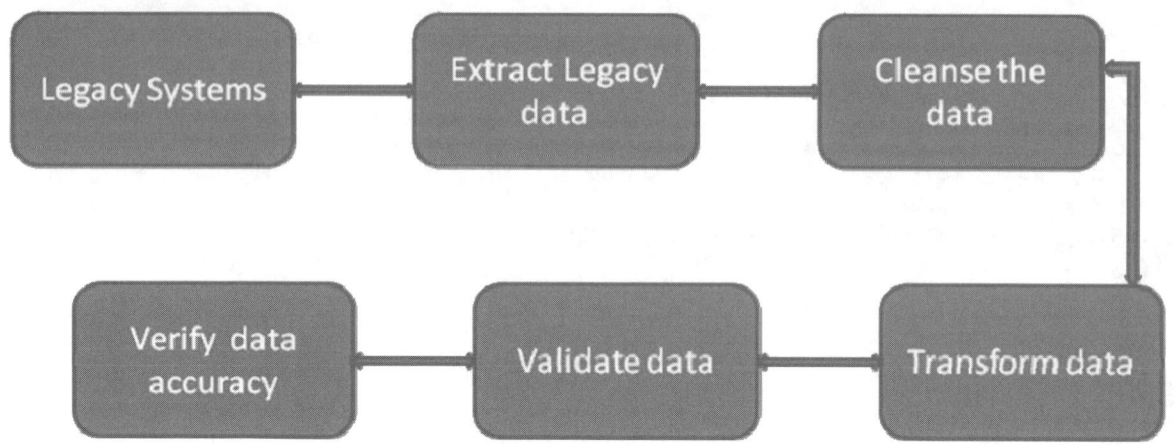

Stage 1: Planning and Scoping

The scope of the data migration must be defined in writing and the appropriate funds should be allocated to it. Different classes of business resources must also be allocated. These resources are: project management resources, technical resources, hardware resources and software resources. Data migration also requires a thorough understanding of the organization's business model, and it is also important to understand the data requirements for the business.

Customers typically do not realize the amount of time and effort needed for data migration. This could lead to having unreasonable expectations for project requirements and scheduling. Project expectations must be set in the beginning with customers so as to prepare them to make the necessary decisions and tradeoffs throughout the migration process.

> **Customers don't often realize the full time and effort needed for data migration. Set project expectations and timelines before the start of the project.**

The project plan for the data migration process should outline the below activities:
- Schedule set for the migration.
- Whether the migration process is phased.
- Description of the legacy and destination data.
- Identify resources and responsibilities.
- Description of the impact of data migration on the business.
- Identify data mapping requirements.
- Describe data management methodology.
- Plan for data verification and testing.

- Establish a contingency plan.

Stage 2: Extracting Data from the Legacy System
Extracting data is the first step in the actual data migration process. The main aim of Extraction is to convert the legacy data into a format which facilitates subsequent stages of cleansing and transformation processing.

Stage 3: Cleansing the Data
Cleansing is needed in order to repair corrupt data, remove duplicate records, and repair/remove invalid records from the extracted legacy data. This encourages consistency in the data across the enterprise. The cleansing process also involves updating the remaining data to fit the format requirements for the new system.

Although portions can be automated, data cleansing is mostly a manual process. Once data inconsistencies and other issues have been identified, you can use customized tools for further cleansing steps, such as updating the data to the format requirements for the new system.

Stage 4: Transform Data
Legacy data has to be transformed to fit the business needs of the new system. In this phase, a series of rules or functions are applied to the extracted data intended to derive the data to be loaded into the new system.

Stage 5: Validate Migrated Data
Migrated data from the legacy system has to be validated against the business rules of the new system. This can be done by applying the different validation methodologies used in the industry. For example, if you are migrating from some other database like the SQL Server then you can use "Oracle Database Migration Verifier" to validate the structural and data integrity of your migrated Oracle database.

Stage 6: Verify Accuracy of the Migrated Data
Although the data has been validated throughout the migration process, final pass verification has to be performed to ensure the accuracy of the migrated data. This verification process is best performed by a group of users familiar with the data and who can evaluate the data in a controlled setting.

5.4.2 Tools for Data Migration

There are a variety of tools available to use in the data migration process by both advanced and beginner developers. These tools are contingent upon the compatibility of the data environment and platform dependent.

ETL Tools

Extract, Transform, and Load (ETL) Tools are important because they facilitate the way in which data will be migrated to a database. ETL is also used for integrating with legacy systems

Because ETL tools are too complex to properly develop in house, companies generally buy them to make sure everything runs smoothly. Good ETL tools are able to read different types of file formats and can cross communicate with various relational databases.

There are several ETL tools for the Business Intelligence / Integration field. These tools help maximize the business value of data for analytical and operational needs.

Informatica PowerCenter – A scalable functional programming tool for platform for accessing, reporting, and integrating data.

Ascential Data Stage – Provides OnDemand data integration.

ESS Base Hyperion – Creates analytic and enterprise performance management applications.

Business Objects: Data Integrator – Builds data marts, ODS systems and data warehouses.

SAS Data Integration – Generates data streams, data marts, and data warehouses.

Microsoft Data Transformation Services (DTS) – Aids data movement issues, and automates ETL processes through databases.

Oracle Warehouse Builder – Provides fully integrated relational and dimensional modeling and the management of data and metadata, among other functions.

Pervasive Data Integrator – Web Service data integration and data management solutions.

Cognos Data Manager – Creates dimensional data marts driven from a common framework.

Ab Intio – A single technology platform for data processing applications.

These tools will assist you in your ETL operations to make smoother data transitions. Evaluate the different tools that and make decisions that are right for the scale of your project.

5.4.3 Issues with Data Migration

Data migration projects typically highlight serious data quality issues that, too often, lead to delayed projects and cost overruns. In extreme cases, poor data quality has led to the abandonment of new business initiatives entirely. And, environments where high volumes of data are entered into key business systems are breeding grounds for data quality problems such as lack of user knowledge, absence of stable and robust processes, and missing relationship linkages. The most common issues are:

1. Incomplete Data
Data can be missing at the field, record or even table level. In other cases, data that is entered may pass the data type validation, but not on the application or report level. For instance, if some field values are null and during the migration process these were repaired or validated, reports that run from the new eCommerce suite may or may not be accurate.

2. Duplicate Data
Multiple instances of the same data are big problems during data migration and often can be overlooked. Since the data format is different in each of the duplicate records, even though the information is the same, it is difficult to narrow down and ignore duplicate data records.

3. Data Non-Conformity
Data in the database may become formatted differently over time from individual to individual during the creation of the legacy database structure. Hence, data is not present in the standard format.

eCommerce Best Practices

4. Inconsistent and Inaccurate Data
When merging various systems, the data can lack consistency and represent wrong information. Data deteriorates over time, which can cause a lot of difficulties during migration.

5. Data Integrity
Missing relationship linkages can drastically degrade the quality of data and generate incorrect reporting after completion of the migration.

Prior to developing a conversion plan for data migration, it is worthwhile to research the format and quality of the data presented in the system. This is similar to the requirements gathering phase of any project; listing all the types of data that need to be converted reduces the risk of errors.
To avoid any serious damages that would impact the business operations, there are some best practices that should be followed. First and foremost comes the planning.

5.4.5   Planning

Just like any other process, planning is an important and integral part of Data Migration. As quoted by Winston Churchill, "Those who plan do better than those who do not plan even thou they rarely stick to their plan." In order to ensure successful migration of data for an organization, the key first step is to make a plan.

Planning depends on the size of data to be migrated. Apart from listing the steps to be followed during the planning process, you should also include the following:

Scope and Objectives have to be clearly identified.
This includes review of the two systems involved, especially the compatibility between the two systems. Secondly, you have to determine how the process will be carried out when the current system is running. Phases for the whole migration process also need to be rolled out and approved from the stakeholders of the organization. In addition to this, you also need to come up with a plan to measure the usability of the "new" system after the migration.

Determining the sets of data to be migrated.
This needs to be determined before the actual migration starts. This step will make sure that the people involved in the process have identified what data

needs to be migrated and that everything is present. Also this will help in determining any potential problems pertaining to the network load.

Document all critical success landmarks.
Knowing what to accomplish and what has been accomplished is important for any project. Clearly defining what landmarks to hit will better structure your data migration operations.

Migration requirements need to be clearly defined.
Explicitly state each requirement for the migration process. The requirements provide a frame of reference for the migration and keep everyone involved in the migration on task.

All source data locations / paths should be defined.
It is in good practice to define the data locations for tracking purposes. By doing so, people will know what to look for based on what has been documented.

Compatibility between the old and new device.
This could be classified under the first point but being such an important issue, it should be treated as a whole new point in itself. It is important to seek the compatibility between the two systems in terms of hardware and software. Also, the end users of the two systems play an important part in comparing the two systems for migration purposes.

The strategy needs to be understood by everyone involved.
Everyone needs to be on the same page so that they understand the next steps and the processes that will get them there. Migrations are generally a team effort that effects multiple departments and programming.

Determine the time frame needed to perform the process.
As the plan is phased out and all activities are listed you should also make sure that the timeframe needed for each of them is determined, as this will allow you to make adjustments for any external arrangements and also for any validation to be done later on.

Security issues related to the migration process.
This will describe any conditions (if any) for the data access and who can access it. Also, this will determine if any data needs to be retained and if you need any encryption on any data set. This ensures that the organization will not face any data security related issues in the future.

Assigning resources to the tasks identified during the planning process.
This is an important task as the whole process will be carried out by a team with balanced skill sets. While the process plan is rolled out and the time slots are allotted, the organization should also assign the people involved for various tasks so that there is no confusion of responsibility.

Assumptions, constraints, and risks need to be mitigated.
Anything outside influences that could effect the migration operations needs to be reconciled a.s.a.p. The longer these issues are avoided the more detrimental to the project they become.

When the data gets migrated, one important thing to not overlook is the communication between all the stakeholders. In other words, the people involved in the process and also people at higher levels that will be affected by the process. A pre-migration test is preferable that ensures the process to be initiated will not modify the data after the migration has been completed. This is needed apart from the mandatory post-migration test.

Data migration is a complex process not to be taken lightly, regardless of the circumstances. It is not only important to evaluate the tools necessary to make the migration, but it is equally important to evaluate the team and service providers facilitating it. Make sure you have the protocol to enlist qualified professionals handling this matter because it is too important of a process to cut corners.

5.5 Content Management

5.5.1   Introduction
5.5.2   Do You Need CMS?
5.5.3   How to Find the Right Vendor?
5.5.4   Considerations
5.5.5   Lessons Learned

5.5.1 Introduction

Content Management Systems (CMS) help organizations manage a large collection of text and media assets under a controlled / monitored environment. CMS also supports a collaborative environment with a wide variety of roles and responsibilities. In fact, the evolution of Content Management System started with organizations in the news and media business—its roots are in the journalism industry.

CMS began as an internal homegrown application to manage and organize the content for newspaper, magazines and others in book publishing business. In the mid 90's, Content Management System developed commercially as branding became increasingly important within the market. Over the years content management system evolved to a Web Content Management System due to the technological boom. Since then, as the complexity of web application increased, so have the tools required to manage them. Today's Content Management System has grown to be a better, more effective product that serves large organizations in various stages of content life cycle.

Is Content Management a luxury or key to survival—this is a question retailers pose when they look at whether to adopt one or not. The first step to take when you are looking at a CMS solution is not selecting a vendor but understanding your current processes, strategy of business goals, and defining the technology that best fits their business requirements. Only once the self assessment is complete can you search for a vendor.

5.5.2 Do You Need CMS?

Some of the major benefits of CMS are making quick and easy updates to your web pages, upholding consistency, eliminating the dependency on the IT department for any updates to your web page, and the content managers can

update the content without depending on the technical team, which make updates faster than using the procedural approach.

In organizations where CMS is firmly established, the web page development is much faster because the content owner can create, preview and publish the content with the help of the CMS tool. In a traditional content based application the content owner has to place a request to the deployment team when the content is ready for publishing. With the help of the CMS tool, content owners can create and publish the content by themselves.

Another advantage of a CMS tool is that you do not need to have HTML programming skills to directly format the content that need to be published. Today's commercial CMS tool comes with a WYSIWYG editor, plus the CMS tool allows you to create the several visual template that can be applied automatically to different web pages in your web site. Once the templates are available to the content owner, it is as simple as completing a web based form. With the help of templates and a WYSIWYG editor, content owners can create the content quickly and publish it to the web.

As the demand for the CMS tool grows in the marketplace and across industries, the CMS toolsets get more and more sophisticated to meet the changing needs. For example, end users demanded accountability and the ability to track changes. As a result, CMS tools now let you control the access to users; you can create different roles and assign these roles to user accounts.

As discussed earlier, the CMS tools allow content owners to create the content using templates because they guarantee consistency across the site. This means that content owners will have control over the dynamic content in the site, but the static content of the website are not modified by content owners, e.g. static content being header menu items, footer menu items and other site navigation elements of the web page.

> **CMS tools are essential for larger organizations as opposed to smaller businesses with less content to manage.**

With all the benefits and features packed inside the tools, CMS tools are must haves for organizations of significant size. Organizations can use CMS tools for creating content for a variety of websites ranging from a corporate intranet portal website to a multi-brand / multi-lingual product information website. CMS tools offer greater independence for content owners and for professional looking corporate sites of any organization in any size.

eCommerce Best Practices

5.5.3 How to Find the Right Vendor

Choosing the right CMS tool is always a challenge. In the past many organizations developed homegrown content management tools to meet their specific needs. These tools range from prevailing and expensive enterprise content management tools to low-end open source CMS tools. You should be aware that these tools may not have all the features that your organization's needs, so choosing the right tool for your company is the most important aspect to consider when choosing the right CMS tool in relation to your needs.

CMS tools have evolved from simple template based data management systems to power tools whose principal purpose is to manage the content directly. In addition it offers support for managing content life cycles, page templates, managing workflow, supporting RSS feeds, personalizing content, and managing digital rights.

Given the choice of CMS tools—they range from the very expensive enterprise-wide applications to less expensive online open source tools—how do you decide which is the right tool for you? Look at your company needs—what is your budget? Do you have the proper time allocated? How much support you can get from your IT department for ongoing maintenance? And, what are all the features the CMS tool has to offer?

> **What is your budget? Do you have the proper time allocated? How much ongoing support can you get from IT? What features does the CMS tool have to offer?**

Building a Business Case
First, you should outline the objective of getting a CMS solution within the organizations stated business case. The business case helps simplify the implementation process also as it provides all the key elements that could be problem areas and identifies the variables that affect it. The report should address the following:

- How much design work is done internally, what quantity is outsourced and how frequently.
- Detailing current production roadblocks and fixes being handled. Predict long-term hurdles and expenses involved.

eCommerce Best Practices

- Growth strategies to include eCommerce production plans, implementation timelines, volume standards of print catalog, and outsourcing.

The choice is also there with should you host the product within your environment or outsource it. There is merely a business decision that should be driven by cost and ability to support the required infrastructure. On the whole, retailers should implement a CMS that is congruent with their information and enterprise architecture. Some of the known vendors in the market who support the CMS implementation are listed below:

Interwoven
An expensive CMS tool like Interwoven covers every aspect of content management tool, they are more appropriate for high-traffic web site as the tool offers full service solutions.

ATG Content Administration (CA)
CA tools offer a complete solution for an organizations CMS needs. This includes managing dynamic database driven content, business user interface (WYSIWYG editor), intuitive interface for workflow task and tracking, seamless integration with enterprise system, revision control, role based personalization, better search capabilities, smart deployment options, multilingual support and support for RSS feed.

Vignette
Vignette's enterprise content management solutions offers business users the opportunity to manage web content online, provide native management and delivery of all electronic assets, dynamic content delivery, and it adheres to evolving J2EE, XML, Web Services technologies.

TYPO3
TYPO3 is an open source CMS tool that offers fully flexible and extendable enterprise solutions for web and intranets. It features ready-made interfaces, functions and modules.

5.5.4 Considerations

There are several reasons why content management would be a useful tool to your organization, at the same time there are several reasons why implementing a CMS can be a costly and impairing decision. For example, if

eCommerce Best Practices

the implementation team has the flexibility to adapt the new CMS system, job security related concerns may arise because the tool lacks efficiency, adds complexity and possibly additional responsibility for content creators. Other CMS tools related issues are discussed below:

Personalization Rules

Personalization is the couture web content for an individual or group based on any criteria. These may include user attributes like gender, locale, and age. It could also be based on a user's behavioral information gathered directly or indirectly from a website. Using a personalization server, you can deliver personalized content to your target audience.

There are several vendors available in today's market that offers personalization servers. An ATG Personalization Server can choreograph customer interactions through scenario personalization. Some of the other leading vendors are BEA WebLogic Personalization Server, and IBM WebSphere Portal. These servers use various rules to drive the content to the user based on the predefined rule of creating more relevant user experiences for individuals and segments. Personalization servers operate in rich J2EE technology, offering a range from portals to content management. This means organizations can offer personalized collaboration sites or personalized eCommerce sites.

Complex, Inter-related Database

To support complex content management applications it is required that enterprise database, application servers, web servers and other servers outside an organizations environment operate cohesively for a successful website. However, these things are not necessarily designed to be inter-related, which makes an IT department's infrastructure increase in complexity.

One way to reduce this problem is by implementing an administration tool that will intelligently automate, monitor and analyze the data that is collected and stored in your CMS application. Most industry leading CMS tools come equipped with an administration tool to eliminate this complexity that will help to quickly add new records or can be derived from existing data using web-based forms that are easy to use by business users.

Dynamic Previewing

The dynamic previewing feature is not offered in all commercially available CMS tools; however, this feature is gaining popularity among CMS tool users and vendors. Having a "preview" capability allows more time to market your products because your content is shortened. Content owners can preview the web page with the static content along with dynamic data driven items and rich web UI interfaces in their true state prior to the content actually being deployed to the production server.

Development Environment

Organizations often try to separate the development environment (CMS system) and production environment (content delivery). In this approach, editors and content creators work in the development environment. When the content is reviewed and ready to be deployed, it is published to the production environment. This results in a number of items to consider during setup:

- **Less Expensive Pages are rendered Faster.**
  This approach is definitely an advantage over the runtime ("On-the-Fly") assemblage of the webpage. Runtime page assembly uses a lot of resources for user click, and is quite expensive. When the user clicks on a link on the page, it triggers the logic to construct the page from the content database and stream it to the browsers.
- **Reviewable & Author/Editor Friendly**
  Runtime CMS does not have the preview capability. And, content is published to production to preview in the final form, which is always a risk.
- **Clean Separation**
  Separating the CMS from delivery system will leave the production server accessible to only a few users.
- **Infrastructure Security**
  The most important steps in site security are protecting your production server from known vulnerabilities. Your CMS tool should be able to communicate securely with the production server during the content deployment.
- **Access Control**
  Organizations may find it difficult to select the right CMS tool for the web site that protects content rights. Administrators may want to limit the user from accessing any sensitive content.

Production Deployment

The most widely-known method to move content between environments is via File Transfer Protocol (FTP); however it is not without its challenges. FTP servers seem to be one of the favorite targets for hackers. Also, the packets transmit the user authentication information as clear text makes the FTP file transfer more insecure. Alternatively, you can securely transfer content using VPN, which limits access down to the known address only.

Unfortunately, FTP servers do not guarantee transferring the content accurately and completely. If for some reason your network connection is interrupted, then FTP clients cannot roll back the transaction (roll back to production server's previous state) or guarantee error checking against file corruption. There are tools that have built proprietary extensions to FTP clients in order to make the content transfer as transactional as possible, but you may have to be careful about the robustness of the tool.

Synchronization technologies offer good deployment alternatives and provide the ability to replicate part of the file system to multiple destinations. The main advantage of this technology is that it supports discrete transactional units. In other words, you can roll back to the prior state if there is a problem during content transfer. Synchronization tools use checksum comparisons or similar mechanism to ensure accurate and guaranteed content transfers.

> **Synchronization technologies offer good deployment alternatives and provide the ability to replicate part of the file system to multiple destinations.**

The bottom line is you need to choose the right CMS tool that has a good transport technique in order to transfer your content files with the acceptable level of tolerance and that is capable of production-readiness.

5.5.5 Lessons Learned

Before selecting the right CMS tool for your organization, there are some interesting considerations and lessons to learn. They are discussed below:

Plan Personalization during Selection Process

Personalization certainly requires run-time assembly of dynamic content. Your delivery system cannot push content until it has the knowledge about the person—who the person is, what are his or her interests, etc. You can gather

some of this information explicitly by asking the user to complete a web-based form or implicitly by studying the user behavior on the site using some metrics tools. Either the personalization or the metrics tools will add additional cost to your content management project.

During the CMS tool selection process, ask yourself this question about personalization. Is it mission critical for the business success? Is it going to add value to your website by keeping the users loyal to your web site? Do you really want to change the content on your site based on the user's data? To what level, do you want your site to be personalized for the user? Is it simply a "Welcome back Joe" greeting or a dynamic web page where the content will be different for different set of users based on their user information?

Site Personalization can be complex and costly. You may want to conduct internal due diligence and create a sound approach when choosing the right tool for your organization's CMS need. For example, if your business requires using personalization and you want to keep the customer loyal, you should implement CMS that delivers the personalized dynamic content on your website tailored to your customer's interest and needs.

Customization in Content Management Systems

Given your budget and resource constraints, you might not find all features that you are looking for in one CMS tool. In this event, it is wise to choose the tool that matches closest to your requirement; one that fits within your budget, and customize the tool to meet your business requirement. Most of the CMS tools today, contain an application program interface (API) that can be used to accomplish this.

As a note of caution when you customize the CMS tool, you are exposing yourself to unknown bugs that may be a consequence of customization. This requires additional testing time to make sure the customization is actually helping your organization deploy the content in the way you intended.

Preview to Shorten Updating Cycle

The extraordinary amount of time spent on reworking the published content is where the need for dynamic previewing capabilities originates. In a traditional Content Management System, to fix a minor text error or broken reference URL, takes longer than it should and involves several steps. Ultimately, it cannot proceed without the appropriate IT approval processes completed. Imagine if content owners would like to make changes to their live content. It involves modifying the content, placing a change request to deploy the modified content, getting approval and engaging the deployment team to get it deployed. This results in a significant amount of time and resource straining. When this kind of error happens, it is not intentional. It happens because the content creators do not have any previewing capabilities until the content goes live. If the content creators or owners have preview capabilities, they do not have to wait until the content is published; they can preview the web page in its true state prior to publishing the content. This saves time and money for the organization getting the right content to the market quickly.

When choosing the CMS tool, it is more important to look at whether features like dynamic previewing capability are available, because it has the ability to shorten your update cycle.

Flawless Deployment

One of most overlooked aspect of content deployment is how and when to seamlessly update the live content on the website. This, of course, depends on several factors like the type of changes you make, extent of changes to the content, and your supporting infrastructure. If your content is deployed to multiple servers, how do you ensure the content is published to all servers without affecting the user experience? Every site is different and it is important to consider user experience right from the planning stage. It is critical for large deployments that take several hours to transpire.

There are a number of strategies available to ensure a consistent user experience. You can consider doing deployments in during off-peak hours when the site traffic is less; however, it is important to consider the impact on the international visitors surfing your site given time zone changes. Another strategy to deploy the content is to deploy the content from the bottom-up. For example, if your site shows news headlines on the home page, updating your headlines first without publishing your detail news first would be inappropriate. The solution would be to control the publishing order by publishing the lower level content first.

Incremental updates on an individual page or group of pages of the site is favored. Instead of republishing the entire site, incremental updates saves time during your deployments. For example, complex, highly interactive web sites with rich personalization features should redirect the user to a single server and update the remaining sever with new content. Then, they should stop accepting new sessions on the server with the old content, redirecting new users to the

updated server while those users on an existing session will see the old content until their individual session ends. The final server should be updated to complete the update to all servers.

This enables each publication to be virtualized; meaning every update to the page creates a new version of the site and manages the user session across old and new versions of the site's content. This concept is incorporated in many CMS tools that are available in the marketplace.
Again, choosing the right deployment strategy depends on the nature of your business needs. Select the right strategy that offers a consistent browsing experience to keep your loyal users happy.

Custom Framework

The decision to invest in an expensive CMS tool should be made very carefully. Often organizations overlook their needs; enamored with the idea of total information sharing among the employees. They forget that this is probably not a valuable goal. Information sharing becomes valuable only when the right content is shared with the right audience at right point moment in time. If your CMS tool is not adding value to your organization, it is not worth investing in an expensive tool with fancy features for content management.

Instead of paying a high price for the features that your organization is not going to use, or not required for your business model, select a moderately priced tool with the flexibility to customize features that fit your needs. Adding a custom framework could save some green for you. It is very tricky to maintain a balance between choosing a costly Out-of-the-Box (OOTB) tool versus customizing the mid-range tool.

Conduct your research. Be aware of the organization's needs, and brainstorm your ideas with your team to choose the appropriate toolset. With CMS evolving into a web centric field, it is not more a standard "document management" solution. You need to analyze and look at your product volumes and its constant need for updates from the front end to back end logistics.

eCommerce Best Practices

5.6 Holiday Season

5.6.1 Introduction
5.6.2 What to Implement for the Holiday Season
5.6.3 Testing
5.6.4 Marketing and IT Synchronization

5.6.1 Introduction

The holiday seasons are critical periods for the success of online retailers. The opportunities are greater during this season with some retailers accounting for more than 20% of their annual sales from the holiday season. Catering to the customers' needs and wants during this crucial period is the key to success for most online.

> **Highlight seasonal content, products and categories and put holiday season messages on the front so shoppers are reminded or assured that it is the holiday season.**

Immediate online sales are boosted when holiday shoppers want to save time as well as find the right gifts. Retailers should know that they do not need to overhaul their existing sites completely for the holiday season. However, if you are launching a new site with the intentions of it going live for the holiday season, you need to consider the time it takes to develop, test, integrate and deploy your new site. Typically, a large-scale project can take six to nine months to complete, but the RFP and requirements gathering process should begin months before any development begins.

When the holiday season begins you should lauch a series of key tactics that can be implemented on the existing merchandising and marketing strategies, as well as ensure that they have an optimum customer experience during the holiday season.

5.6.2 What to Implement for the Holiday Season

Clearly retailers need to understand what the best and most effective methods are in order to drive their loyal customers to their site during the holiday season.

eCommerce Best Practices

SEM and email marketing are some of the tools retailers are using to establish a customer base and cultivate profitable and lasting relationships. This is different from the traditional price-based promotions as a method to instill customer loyalty. Holiday season campaigns are measured by defining the more lucrative target segments and identifying your customers across demographics to ensure successful sales according to their needs or interests. Some of the strategies to consider when implementing an effective holiday season plan include:

Countdown for Shoppers

Time-tailored promotions and specials help shoppers count down to the holiday season. Campaigns like "12 days for Christmas" encourage shoppers to start planning and finding the right gift before the holiday and these early purchases help retailers plan more repeat buyers' incentives.

> **Limited time offers help shoppers plan and purchase their gifts on or before the season. Good for the eager shoppers.**

Give shoppers a sense of urgency with taglines like "Offer of the day" or "Gone tomorrow." These prompt the shoppers to make their purchase decision today knowing that tomorrow it might not be the same offer.

Special Offers

There is definitely increased traffic to your site during the holiday season, but equally inviting would be if you throw in some offers like free shipping with a certain amount of purchase, free gifts, or even special repeat buyer discount to fuel repeat purchases from existing customers. These special offers enhance your customer loyalty relationships and ensure a consistent flow of new and old customers.

> **Repeat buyer discounts are ideal during the holiday season as it fuels repeat purchases from existing customers who have already brought products.**

Free shipping is vital as holiday shoppers are bound to purchase more when they know they do not have to cover any shipping costs. Most retailers offer this but you should also have legal implications and regulations to cover geographic shipping issues and costs. Always add a buffer day to the delivery deadlines when your logistics team says they can accept orders until 'X' day of the month. This is critical as during last minute sales, you do not want your overall customer experience to falter due to not having the buffer day to fall on.

Site Features

Personalized promotions, suggested items, gift idea centers, and live chats are some site features retailers use during holiday season promotions. These are used primarily to generate revenue from loyal customers. Most of these features should be personalized according to their click stream behavior of the customers in order to get maximum results.

Highlight seasonal content, products and categories and put the holiday season front and center so that shoppers can find, choose and buy gifts easily. Taglines like "Shop for our exclusive gifts" get the message across clearly — recipients are unlikely to get these gifts anywhere else.

Last Minute Gifts

Focusing on last minute gifts is clearly a vital marketing strategy that can acquire shoppers during the holiday countdown campaign. Last minute solutions can include anything from gift items, promoting gift cards, gift certificates or downloads, free shipping, etc. If retailers add strategies like pre-wrapped, pre-selected and pre-packaged gifts in themed baskets then customers are more favorable to make a purchase from your site. Gift giving is a viral community experience; every gift purchased online impacts the giver and receiver.

> **Impulse buys are commons in the retail industry, offline or online. Creating a space for last minute gifts will drive additional sales.**

Email Marketing

One of the most widely popular and important demand-generating tool for holiday success is email marketing. To a lesser extent, search engine marketing is also used. Email marketing has to be targeted and relevant to produce the desired results. Retailers choose strategies that are embedded within the email marketing and target their customers effectively. It is cost effective and results are guaranteed and tracked for future strategies. Also equally, it is effective in building customer relationships.

eCommerce Best Practices

Retailers can use gifts giveaways with frequency twice or three times a week with featured product discounts or free gift-wrap or shipping offers. When the season approaches, daily reminders and gift of the day promotions can speed up the purchasing process. Using email segmentation and personalized emails should create offers by the channel of origin or purchase, frequency and spending history.

Shoppers on a budget can be sent emails about gift ideas in their price range. On the other hand, top-end buyers can be offered free gift wrapping or premium shipping. Multi-channel merchants can encourage in-store shoppers to shop online and vice-versa with targeted offers like free shipping for online purchases or free in-store gift cards to in-store shoppers purchasing more than a specified amount worth of items.

Shopping Guides

Interactive and useful guides help the shopper during the busy holiday season to narrow their choices and give more time to shop for other items as well. Gift finders are an example. Some guides even base the decision on whether your recipient was "naughty" or "nice".

Retailers need to remember that during the holidays, people coming to their site may not necessarily be your typical customer. Instead they could be buying gifts for other people that are your typical customers. Thus, your site needs to present a different "point of view" to give newcomers a little extra guidance.

Customer Relationships

Though email marketing is effective when it comes to creating personalized targeted campaigns, retailers also need to look at other avenues to create and encourage more purchase interests like: special offers for customers such as free delivery or easy returns, special events for customers, etc.

> **Foster customer relationships by making memorable and personal associations with them through special events or offers.**

Evaluate and Compare Metrics

Retailers should rely on response and activity-based metrics like click-thru, page and product views to measure the generated demand. A huge amount of value is gained from response and activity data and you can use these to identify what is exactly needed for the upcoming holiday season sales and promotions.

This data can also serve as basis for relationship marketing. You can use this to target their customers based on their interests.

Post Holiday Season

With every gift received you now have a new potential long-term customer. Retailers should not end their season as the holiday comes to a close. Instead use the profits and cash on incentive offers like extending return policies beyond the holiday season. These ideas are also viewed as additional brand contact points.

5.6.3 Testing

Because things always manage to fall through the crack and there always is room for improvements, it should be your job/goal to fix last year's and previous years' mistakes. Retailers can adopt operational audits that help in fact finding and determining areas of concern to increase performance and productivity from previous years. More effective use of distribution centers, capacity of orders processed, streamlined workflow and higher profits/lower costs are some goals to achieve prior to the holiday season.

Research Reports
These reports need to come from three areas – Internal, Returns and CSR. Internal reports can consist of Quality Assurance, Pick-up, Packing, Inventory Center, and Shipping data in which detail activities about efficiency and service for each section. Weekly reports show which sections have reached their desired level during any given time period (last season or last week), and gives you the insight to control and apply those successful strategies well before the start of the season. Apart from these internal reports, retailers also should look at Return reports from the customers' end. This should be segmented by product, price, vendor, time of return and shopper association. Reasons for returns, i.e. duplicate items, late delivery, wrong item by color, size or quantity, damaged item or shipped to wrong address, should all be marked and noted if a pattern is discovered.

Customer Service Representative Reports provide a valuable source of information about your customers' complaints, queries, as these are not

eCommerce Best Practices

just limited to operations and fulfillment issues. Unfulfilled promises made last season to customers are redeemable should a repeat occurrence is prevented. With respect to CSR, retailers can also conduct call monitoring systems which yield valuable customer information, shows whether you have had effective CSR training, and whether or not customers are happy with their service.

Internal affairs operations should be implemented to ensure customer service quality control. It is in best practice to have an internal resource perform random tests of customer service effectiveness. They should place orders and pose as customers to see how their CSRs respond and take action, look at how fast the fulfillment center works and if their delivery was on time. Once these varied reporting systems are in place and the necessary information is gathered, the next step is to plan benchmark comparisons.

Comparisons and Benchmarks

Once all the initial research reports are complete, the next step is to compare the standards of service with the actual performance. Also figures related to goals and projections should be compared with other Multi-Channel business vendors such as yours to help evaluate your own individual performance. Find vendors with similarities in regards to products sold, price, business model, operations or shipping modes. Retailers should survey fulfillment centers as their effectiveness is vital during your holiday season sale. Work flow of processes from one department to another, space requirement variations during peak and non-peak periods, staff responsibilities and roles, and overtime labor rates are some areas to focus on when managing the fulfillment centers.

Action Plan

Once all the reports are gathered and compared with benchmarks, certain patterns will emerge that can help you identify areas of improvement, stability and success for the future. The final step is to create an action plan and test which one yields greater results over your previous season sale. Always flesh out the most optimum plan and keep everyone involved. Changes introduced gradually are bound to be more effective than one single massive overhaul.

But how do you know if your site can handle the increased traffic? What if your season's marketing strategies turn out a big success and you have no idea of

eCommerce Best Practices

the incoming shoppers. Retailers need to test the key sections of their site expected to make the biggest splash.

5.6.4 Marketing and IT Synchronization

Retailers can never be too prepared for the onslaught of traffic they will experience during holiday season. Your brand's promotions, emails, and other marketing strategies drive customers to your site, and they seem to collectively wait for a certain time period to flood your site and start their buying process. If your site, server and other IT infrastructure combinations are not synchronized, you will most likely be faced with a major economic meltdown, with all your marketing strategies at waste due to a lack of IT support. Weak IT performance can leave you guessing at how many customers you lost this season and how many will not return for the next season. Retailers need to encourage their IT and Marketing departments to create, work together and have self contained slots on their site that can be changed on the fly without interrupting the shopper at the same time.

Free shipping is a great promotion during holiday season but you should look at a potential burn out issue if you are not able to cater to an influx of unexpected orders during the duration of the season. All promotional landing pages need to be tested to make sure that shoppers are clicking through the relevant pages and getting the appropriate messages. Landing pages with different product categorizations also need to be tested before hand so that products are arranged the way a shopper likes to shop. The holiday season might run only for a short period out of the year, but the amount of traffic data that gets generated in during that season is significant enough to warrant IT upgrades or extra support. Testing should always be done outside the peak season with enough buffer period in-between to undergo changes. Synchronizing your marketing strategies with capable IT support is absolutely critical.

Every retailer should be able to look forward to the holiday season with the confidence that every transaction, interaction or operation goes off without a hitch. In order to do so, you must learn from your mistakes from previous seasons and test all upgrades or changes you make before the holiday season begins. Clear instructions, well categorized product displays, and congruous landing pages will help minimize challenges on both you and your customers' behalves. With promotional tactics applied well ahead of the season, retailers can plan and get a feed on the oncoming traffic. You can start their plan for backend logistics and anticipate the rush. Minor improvements can make a major change.

5.7 Multi-Channel

5.7.1   Introduction
5.7.2   Marketing for Multiple Channels
5.7.3   Consistency in Multi-Channel
5.7.4   Core Competencies Requirement
5.7.5   Multi-Channel Integration
5.7.6   Customer Servicing in Multi-Channel

5.7.1 Introduction

Retailers that once relied on their brick and mortar store and catalogs as merchandising avenues, can now offer a multi-channel approach to their business model. Online stores and mobile devoces added new channels for retailers looking for market separation. By utilizing stores, websites, and catalogs retailers optimize annual spending, business transactions and long-term loyalty from customers. Multi-channel success requires applying all your efforts in a variety of marketing channels to bring in and retain customers.

Creating a balance of marketing efforts between each channel can be complicated if you do not understand your target. Many customers prefer one channel over another, and it is important to determine what percentage of investments will go to each channel. Cross-channel marketing should also be put into consideration when putting a strategy in place, for example, driving people from the catalogs to the eCommerce site.

Some retailers think that promoting alternative channels is more detrimental because it confuses the audience as to where they can go to shop, however if you promote each channel you will increase the number of shoppers. Many consumers combine online and offline channels to do their shopping — combinations including catalogs, in-store visits and the eCommerce site.

There are also multiple ways to reach out to customers along the multi-channels. You can send emails with an embedded link to the website or send SMS messages to mobile devices about nearby stores with respect to their location at that time.

Retail customers that make purchases through mutliple channels tend to spend, on average, a higher percentage per year than those using a single

eCommerce Best Practices

channel. This is attributed to the added benefits of using Mutli-Channel. Because there are more channels to communicate with the customer, they experience the following:

- More personalized service and fast access to information. Online stores and mobile devices allows for more frequent messaging to customers.

- Greater access to promotions.

- Focused messsaging and actionable information.

You should have Multi-Channel marketing on top of your to-do list, if it is not there already. Surprisingly, only a fraction of retailers segment customers according to multiple channel usage and marketing.

5.7.2 Operations in Multiple Channels

The operations between Multi-Channels need to be focused on getting the most out of each channel. In order to know the value of the channels, there has to be significant data collected and analyzed through each channel.

- The data collected at every point of customer interaction should be standardized to measure the effectiveness of all channels.

- The customer information collected through every channel should be fed back during the customer's next purchase pattern.

- The buying experience should being consistent across all channels. The products offered online should be the same, if not greater, than offline methods.

- Customer policies should be maintained across all channels—return policies, sale credits, fulfillments should be standardized regardless of the channel being used.

> **Customer policies have to be maintained regardless of the channel. Universal, quality service has to be provided.**

- Make sure your staff is informed about all channel activities so that they are prepared to help customers regardless of the channel being used.

eCommerce Best Practices

- All promotional offers from each channel needs to be accurately tracked with respect to conversion rates or clickthrus.

- Measure cross-channel strategies for different demographics. Create promotions that span various channels.

You will likely find different patterns and preferences for different channels. Keep testing each channel's effectiveness by developing cross-channel promotions and tracking the results.

5.7.3 Consistency in Multi-Channel

In order to maintain a lasting relationship with your customers, the branding, products and overall design must be consistent throughout the channels. This consistency should extend to sales promotions and offers as well. You have to be careful not to introduce strategies that target the same buyer who purchased a product from your catalog a week earlier, with an email offering the same product online with a discounted price.

> **Multi-Channeling bolsters customer loyalty and repeat business, but only if you maintain brand, product and presentation consistency.**

Promotions are a bit more complicated for Multi-channeling than for single channels. If you are going to offer a channel-specific promotion that cannot be offered in other channels, then you need to clearly state that on the promotion. Also, if your promotion has restrictions on them, i.e. product limitations or pricing minimums, it should clearly specify whether it relates to all channels or not.

Customers are particular when they shop and want what they want, when they want and how they want it. If you do not offer multi-channels when shopping, you lose out on a large, valuable customer base that may think your brand is ignoring their shopping needs. Single channel shopping puts limitations on the ways you can receive revenue. The advantage of Multi-Channeling is that is supports customer loyalty and repeat business.

5.7.4 Core Competency

To succeed in Multi-Channel environment, retailers must first integrate the core competencies. Some of the standard aspects are listed below:

eCommerce Best Practices

Creative

How does your branding tie into your creative Multi-Channel marketing? Strong branding suggests consistency. Making sure your design aesthetic is shown across all channels is important. Familiarity brings about longevity—the longer a customer is exposed to a consistent brand image the more time he or she has to become attached to it. The key here is reinforcement. Multi-Channels reinforce the ideal that your company is dedicated to satisfying their customers.

The imagery should remain the same in terms of design aspects. Store signage, retail advertising, catalog design, and every Internet marketing effort should incorporate the same theme or motif. Color palettes, photographs and illustrations should be uniform with fonts consistent in style and size.

You should make certain that offers and promotions consistently reflect the brand and positioning — you do not want to confuse customers. Stores can offer catalogs with every purchase or checkout counters. Online kiosks provide an added advantage for customers who do not want to walk through your store.

Merchandising

Keeping track of merchandising through each channel is also harder than operations running on a single channel. Retailers need to have the flexibility to mark down products based on inventories or seasonality. Because print takes significantly longer to update than online media, inventories and prices may not be accurate for catalogs.

> **Inform CSRs of all the promotions in their respective channels so that customer inquiries can be attended to successfully.**

Also, each channel has its limitations. Print catalogs and mobile devices, for example, cannot display every product and color variation due to page layout restrictions, whereas that would not be a problem online.

eCommerce sites have more flexibility for merchandising, and can make the shopping experience feel as life like as it is in the store.

eCommerce Best Practices

5.7.5 Customer Service in Multichannel

Rapidly evolving technologies have created more ways customer service representative (CSRs) can engage with customers. Though there are still the traditional methods for customer service like in-store employees or call centers, more recently technology has advanced to include emails, click-to-chat and click-to-call, and even text messaging. These leverage the younger demographics' preference for using indirect channels of communication.

> **CSRs should be informed about all promos and catalog launches before the customer so that they can be prepared to assist them.**

However, because of the popularity of these communication channels, finding qualified CSRs is a little harder. Representatives that are wonderful with customers on the phone may not necessarily be communicative via emails or chats. Also, many telephone exchanges end with confirmations of orders via email. Being able to jump from communication channels is important.

Personal attention to each customer is a part of good service, however, juggling personal attention to several people simultaneously is harder than you might expect especially because products' item numbers may vary from print catalog to online.

So how do you make sure the exchanges between CSRs and customers is smooth when you offer Multi-Channel ordering? Below are some standard considerations:

- CSRs should be prepared. You need to make sure that whenever any new promotions or catalogs are brought to customers, the CSRs also should always be know about them before their customers get word.

- Increase salaries for CSRs that are involved in more complex communication channels. Balancing customer interactions throughout different mediums is not easy and should be rewarded properly.

- Train CSRs, old and new, on the types of communication mediums you have and what the best practices are for engaging in multiple conversations at once.

eCommerce Best Practices

- Monitor CSR interactions to find out what methods work versus methods that are less successful. Quality assurance processes need to be in place in order form a solid customer service operation.

Once a fairly intricate process to manage, Multi-Channeling is now easier to operate with new eCommerce systems and applications. Consistency can be maintained for each channel and updated product or customer information can be generated from channel to channel. Different customers may prefer one channel over the other or perhaps they like utilizing a combination of them for purchasing or customer service, but the important thing is that they are given the option. Many retailers have a website for name recognition, but are limiting their profits by not offering an eCommerce capability. Transitioning into offering an online market channel is worth the effort because the rewards are often greater work put into it.

eCommerce Best Practices

5.8 ROI Analysis / Cost Analysis

5.8.1 Introduction
5.8.2 Organizational ROI Mindset
5.8.3 ROI in Analytics Tools
5.8.4 ROI in Behavioral Targeting

5.8.1 Introduction

As a critical metric for businesses, ROI computations are often difficult to accurately calculate sometimes not granular enough to be meaningful and comprehensive to the broader organization. To ensure that projected ROI can be accurately realized, you need to organize your efforts around key areas, the organizational mindset, optimization processes, key metrics and technology.

Most established retailers predetermine how ROI will be recognized prior to implementing tools and measuring it via mainstream methods like DCF(Discounted Cash Flow) and NPV(New Present Value) calculation models. However, other variables to consider when projecting value should include the ROI of online and also offline efforts. ROI is driven by the role that a site plays in selling a particular class of products. These may include the following:

> **Organizational mindset, optimization processes, key metrics and technology are key areas for ROI success.**

- A replenishment goods website which sells health and beauty products where customers are predisposed to simply reorder.
- A searched goods websites for products like consumer electronics and furniture where purchase is a lengthy procedure requiring more information or technology.
- A convenience goods websites which include apparel and toys for which consumers shop and spend more.

5.8.2 Organizational ROI Mindset

How do you promote a cross-departmental ROI mindset and mature from an activity-based cost savings model to a result-driven revenue generating online retailer? There are many suggested options from marketers but the standard methods are to build performance metrics, host weekly data collection and create staff accountability for the measurable results.

Performance data should start by establishing scorecard metrics that can be captured and grouped by online campaign (i.e. key performance indicators or KPIs). Periodic administrative meetings allow management to review, discuss, and reset the direction you want to go in based on near term results. Ensuring the staff is accountable requires setting personal incentives within individual performance reviews and reward the abilities to meet or exceed the established targets. Some of the processes to initiate measurement within the ROI techniques are:

Reporting

Retailers need to focus their reporting on key data based on their management-driven, operational objectives. For example, if retailers set the primary objective as their eCommerce, then standard KPIs would measure the average order size and browsers to buyer conversion rates. Similarly if the site is geared towards lead/sales generation rather than direct purchasing, then a different set of data should be measured.

Decision Support Systems & Conclusions

You have to analyze data to identify trends and pinpoint areas that needs improvement. For example, shopping cart conversion funnels help retailers determine what percentage of visitors bail out when they get to the shipping information page. Examining the abandonment paths of the visitors would help further the immediate indication of what needs to be fixed to ensure customers are moving towards conversion stages.

> **Pinpoint key areas that need improvement and provide immediate indication of what needs to be implanted for conversion stages.**

After analyzing the data, you might not be too sure of what conclusion to draw from the data and when to decide to make changes. Sometimes all major changes are done at once and results may not be too reassuring from later data that is procured. Multiple changes make it difficult for you to know which specific changes had the greatest impact on performance. Action follows decisions; you cannot improve results if you do not act upon what the data tells you.

5.8.3 ROI in Analytic Tools

Many retailers possess high performance analytic tools but sometimes they find that these tools are not providing what is required and expected. Higher-end solutions like Web Trends, Omniture, or Coremetrics have capabilities which may outrun what a basic retailer requires from their sites. No tool would automatically improve site performance right out of the box. You have to select and configure them to your system. Some issues retailers have with their unsuccessful data metrics could be a result of:

> Get the right analytic tool that will deliver the appropriate ROI analysis based on what you need and not what the tool can provide.

Data Collection Has No Methodology

Some of the tools provide more than 4-6 million different views of your site's website data. You need to implement a standard methodology to analyze the data based on overall site goals, and identify opportunities and test solutions using analytics. You need to have a designated team to process the data and move the information forward.

Focus on KPIs

The data has to be clearly defined and prioritized before the KPIs are received to get the maximum data results. Focus is lost when the data is not clear and there is no priority level defined.

Poor Implementation Tools

Retailers need to ensure that the set tools that provide information on key data are implemented and configured correctly. Otherwise, data collection would not be accurate and misguided patterns will arise.

Accuracy Issues

If incorrect or incomplete data occurs due to poor tool implementations or less focus, then you should not act on them. Accuracy issues occur when you mistakenly adopt the incomplete data.

Poor Communication

Retailers need to ensure that people who control the tools communicate with those responsible for the website.

Tool Training is Must

Each tool is unique in how it presents its data and lets you to navigate and acquire the desired measurements. The analysis team should be trained or have a basic understanding of how to exploit the tool and know where to look and which figures to focus their attention on.

5.8.4 ROI in Behavioral Targeting

The value of behavioral targeting in eCommerce is unquestionable. Every shopper has their own unique way of processing the messages and products shown to them online. By tracking what their interests are and how they shop, you can more effectively drive them to checkout. This is done by up-selling or cross-selling and personalizing your messages to them.

> **Behavioral targeting ROI ensures effectiveness and efficiency and optimizes cross and up sell opportunities.**

Every customer has his or her path to checkout. The same behavioral targeting methods for one person might not work for the next person. ROI is a result of identifying the customer's shopping pattern and directing marketing efforts to fit that pattern.

Key metrics to look for in behavioral targeting tools are measurements like key interaction rates, conversions, content paths, website visitation frequency. The visit quality index looks into to the site's highest value content and whether or not visitors were driven to it.

In an industry where technology constantly changes, the stress of ROI is always expected, if not demanded. One of the best tools to have in eCommerce is a behavioral tracking tool because the information collected is invaluable to creating long-term brand relationships with every customer.

eCommerce Best Practices

5.9 Rough Order Of Magnitude (ROM) Estimation

  5.9.1   Introduction
  5.9.2   What is Rough Order of Magnitude (ROM)?
  5.9.3   Whose Fault is it Anyway?
  5.9.4   A Methodology for Estimation
  5.9.5   Analysis of the Methodology
  5.9.6   Best and Worst Practices
  5.9.7   How to Use a ROM Estimate

5.9.1 Introduction

Estimation is a task not unique to eCommerce, software engineering or to other engineering endeavors at large. However, just as estimation approaches for software projects differ in particulars from ones used in other engineering disciplines, eCommerce projects differ enough from general software projects to warrant variance and attention to particular details that are different in an ecommerce implementation, if not in kind, in importance.

Businesses rarely undertake a project that is thought to be too expensive or too lengthy, and any project that substantially exceeds its original estimated cost or duration is either cancelled, deemed a failure, or both. It is therefore crucial for a project to be deemed successful, that its implementation matches roughly with the estimate that is invariably required for a decision on its undertaking.

Given its importance as the ruler by which the success or failure of many projects is measured against, it is surprising what little resources and importance are dedicated to the task of estimation by software organizations. It is not uncommon for the sales department to adjust an estimate to match the perceived expectation of the client or stakeholder. Often, estimation is performed by someone who is not an engineer and/or a team member that has little knowledge of what is entailed in the implementation of what's being estimated. There is rarely a documented process and guideline to follow when estimating a task.

Accurate ROM estimates are not only possible, but predictable if approached methodically and using engineering principles. This chapter explores best practices to achieve accurate ROM estimates in software projects and highlights a few nuances particular to eCommerce.

eCommerce Best Practices

5.9.2 What is Rough Order of Magnitude (ROM)?

Rough Order of Magnitude means different things in different contexts. In a mathematical context, an order of magnitude typically means a number's nearest power of ten. However in the context of project estimation, this definition would be too rough to make it usable. Few stakeholders would deem a project that took ten times as long to implement or cost ten times as much as originally estimated to have met the estimate.

Also, in the context of estimation, the "roughness" is itself a function of the magnitude. That is, a project estimated to take a week to complete but which ends up taking two, may very well be considered to be within the estimate. However, a three-year project which takes six years to complete would likely be considered a failure. Engineers must be aware of this fact when estimating and managers must be aware of it when planning. This is not a mere curiosity, but a factor which plays into the methodology expounded below.

So in the context of estimation, the acceptable roughness of a ROM estimate is somewhat subjective, but more importantly, the coarseness or roughness depends on the magnitude itself — the former diminishing as the latter increases, requiring greater effort to produce.

5.9.3 Whose Fault is it anyway?

You would think that when a project takes two or three times to complete as originally estimated, the estimate would be deemed faulty. Wrong or right, that is not always the case. It is not always easy to determine what or who is at fault. Often the implementers, project managers and others who may or may not have had input into the estimate are faulted with their inability to produced the estimated results, overlooking scope creep, etc. If the estimator omitted large tasks from the estimate then that is where the blame will land. But more often than not, there are no such blatant omissions, yet the failure is in the estimation. After all, a good estimate will take into account scope-creep, the variance in productivity of different implementers, and in the case of eCommerce projects the likelihood that some requirements will change during implementation and shortly after what had been considered completion of the project.

> Failure of estimation results from omission, but it is likely insidious rather than blatant.

So a failure of estimation results from omission, but it is not necessarily blatant and discernable, but systematic and insidious. Later on we will explore how insufficient decomposition of the tasks to be estimated and performed results in failure due to this systematic omission from the estimate.

5.9.4   A Methodology for Estimation

The estimation technique expounded here is intended to be used prior to a detailed analysis and requirement definition phase where only high-level requirements have been identified. It is an accurate tool for measuring order of magnitude and has been employed in a number of projects with overall excellent results.

Classification of Tasks for the Purpose of Estimation

For the purpose of estimation, tasks fall into two classes: those that are a directly associated with the implementation of a specific requirement, and those tasks that require the successful execution of the project overall. This second class includes tasks such as project management, development environment setup, configuration control, defect tracking, knowledge transfer, etc. The estimates for these are typically a function of the magnitude of the project at large.

It is essential that these two types of tasks be treated differently when estimating them, as the failure of many estimates results from the treatment of the first group in the same manner as the second.

To that end, there should be no attempt to organize and compile requirements in logical or functional groupings. The requirement should be viewed in isolation, as if all other requirements were already implemented. Even more important, one should not try to estimate organizational groupings of activities. Estimation is essentially different than planning and implementation. Planning, and to a certain extent implementation, require a global view of interrelated activities, while successful estimation depends upon the approach that views each activity in isolation.

For example, when a project plan emerges, there will likely be a "Database Modeling and Design" activity. On the other hand, correctly defined functional requirements will never call for such a thing. Instead, the requirements will call for the need to persist user profile information, to persist order history, to maintain a product catalog etc. Each of these requirements may require some database modeling and schema design which may make sense to address collectively, but it is essential for the accuracy of the estimate, that the effort be estimated within each of the functional requirements that call for it, and not in some "schema design and modeling" activity that groups them together. The reason for this counter-intuitive approach is that the effort required for the activity is a function of the number of requirements that call for it, and it is very difficult to envision, and therefore, properly estimate all the requirements that, for example, require database modeling.

Relative Unimportance of Economies of Scale

It may be argued that estimation using this approach discounts the efficiencies of scale. Gathering all the requirements first and then undertaking the schema design and modeling is much more efficient than doing it for each requirement. But, the technique's analysis indicates that economies of scale are much less important in an accurate assessment of magnitude, which is the objective, than accounting for all the required activities.

> **Accurate assessment of magnitude relies much more on accounting for all required activities than on economies of scale.**

Horizontal vs. Vertical Decomposition

A common failure of estimation efforts stems from attempting to estimate the efforts to implement a requirement at too high a level. For example, an eCommerce website will likely require the implementation of shopping carts—a place where customers may place items for subsequent purchase. It would be a mistake to try and estimate the implementation of this requirement at this level likely resulting in an inaccurate estimate. Instead, the estimator should look to decompose the requirement into detailed requirements that stand a better chance of being accurately estimated, but more importantly, will force analysis of what is entailed for its implementation.

> **Consider the functionality properties first when you are in the decomposition process.**

This decomposition should be along the horizontal or functional dimension instead of the vertical or disciplinary dimension at all levels but the lowest. For example, the shopping cart requirement should be decomposed along the functional dimension into lower-level requirements such as: association of

shopping cart to user, adding items to cart, clearing cart, removing items from cart, re-pricing cart, persisting cart, etc., instead of along disciplinary lines such as: analysis, design, implementation, test, deployment, documentation, etc. If each of the lower-level requirements is granular enough not to be further decomposed, then the vertical decomposition is applied, so that for each, an estimate for its analysis, design, implementation, etc, will be produced.

This is not the only break-down possible, and finer granularity is possible along the vertical decomposition as well. For example, in a multi-tier system, implementation may be broken into front-end development, middle-tier development, and back-end development.

Finally, at least some of these elements should be estimated within a range rather than absolutely. The variance in the range should reflect the uncertainty of the estimate. Once again, it is important for the uncertainty to be expressed at a granular level and allow the collective uncertainty to be derived from it.

5.9.5 Analysis of the Methodology

The methodology's effectiveness was assessed by analyzing the result of six (6) projects. The assessment revealed that the estimate for two (2) of the projects correlated less than 70% to actual, while the other four (4) correlated in excess of 90%, ranging from 92% to 97%. Further analysis indicated that the correlation was directly proportional to the number of components into which each requirement was subdivided for estimation. That is, the greater the granularity achieved in the decomposition of a requirement into its corresponding activities, the greater the correlation between its estimate and the actual time that it took to implement it.

The analysis also revealed that the accuracy between the actual time to complete the smallest identifiable activity and its estimate did not contribute significantly to the overall accuracy. So even when the estimate for an activity was grossly inaccurate, or even when an activity identified did not even turn out to be required, the overall estimate was not affected. However, when not enough activities were associated to, and estimated for a given high-level requirement, the estimate for the particular requirement, as well as the overall estimate did show a corresponding decline in accuracy.

5.9.6 Best & Worst Practices

In summary, the methodology depends on these best-practices which if adopted and applied rigorously will maximize the accuracy and reliability of the ROM estimate:

- Divide and conquer; resist the temptation to estimate without decomposition.
- Decompose along the functional dimension until no further decomposition is warranted. Resist decomposing high-level requirements along disciplinary lines such as analysis, design, implementation, test, etc.
- Estimate each requirement in isolation resisting the urge to group them.
- Express estimates for low-level requirements as a range that reflects the uncertainty of the estimate.

5.9.7 How to Use a ROM Estimate

An estimate is typically expressed in terms of Person Days (PD) and in its simplest and most efficient way, indicates the number of days it would take a person to complete the task. It bears emphasizing that efficiency begins to diminish as each task is assigned to more than one person. In spite of well-documented fallacies regarding allocation of additional human resources to shorten the time required to accomplish a task, going back to 1975 and the publication of *The Mythical Man-Month: Essays on Software Engineering*, managers continue to insist in the practice.

Of course, practicality dictates that a number of implementers rather than a single person participate in the implementation of a project, but as more and more people participate there are increasingly diminishing returns. Since allocation of resources and task assignment typically takes place overtime after the production of the ROM estimate, adjustments should be made to compensate for the inefficiencies that arise from the allocation of multiple resources and the communication and managerial overhead incurred.

eCommerce Best Practices

5.10 PCI Compliance

    5.10.1   Introduction
    5.10.2   Risks
    5.10.3   Requirements
    5.10.4   Evaluation Processes
    5.10.4   Validation

5.10.1 Introduction

PCI Compliance (Payment Card Industry) has been hot button security topic in the recent years especially due to the number of credit cards that has been stolen from some big name retailers. This theft left a large number of users vulnerable to identify theft and fraudulent charges as well as cost credit companies millions in issuing replacement cards and other related issues.

The PCI Security Standards Council came about as a result of ongoing security breaches. This council was formed by Visa, Master Card, American Express, Discover, JCB to manage the PCI Data Security Standard (DSS). Note that strictly speaking, you only need to be PCI Compliant if you store or transmit the primary card number. However, the guidelines provided by PCI DSS can be used for any sensitive information that you need to protect in your company.

5.10.2 Risks

The benefits of making your system secure and hacker proof have already been mentioned earlier in this chapter. While there seems to be a few benefits, failure to take these preventative measures can lead to some very serious drawbacks if your company's system is compromised. A hacked system can lead to:

- Bad Press,
- Loss of good will with customers,
- Lost revenue,
- Lawsuits,
- Associated costs in dealing with extra customer queries,
- Associated costs in providing credit monitoring services,
- Cost of investigating the security breach

5.10.3 Requirements

The goal of the PCI DSS is to define a set of rules by which anyone who is storing, processing and transmitting credit card information must comply in order to make sure that the credit card information is not compromised.

In that regard the PCI DSS has twelve requirements that companies must abide by. These are:

1. Install and maintain a firewall configuration to protect cardholder data

    A firewall restricts the traffic coming in and out of a machine or network and in this case the machine which holds the credit card information. This is the basic security measure that should be taken in order to prevent the most rudimentory attempts to steal user information.

2. Don't use vendor-supplied defaults for system passwords and other security parameters

    Failure to change the default authentication credentials is essentially an open door for anyone to come in and steal card information. As basic as this requirement is, it often gets overlooked due to the multitude of products and systems being used in a typical eCommerce site. An administrator might not be familiar or even aware of the different default accounts in a system and may fail to change the passwords on a number of products. A very small amount of research is required to find all default accounts and change them.

3. Protect stored card holder data

    Even if the security of the server is compromised, the credit card information itself should not be stored in a manner which is readable i.e. the credit card number should be encrypted. The CVV and CVV2 number should not be stored at all, however some basic information such as the holder name etc does not need to be encrypted.

4. Encrypt transmission of cardholder data across open, public networks

    The credit card information would require transmission to credit card authentication system or some other third party system for

authentication/credit/debit. The information itself should be encrypted when being send to a third party to prevent someone intercepting and reading the details. It is not enough that the information is just encrypted, but should also be using a strong well recognized encryption algorithm.

5. Use and regularly update anti-virus software

   As the requirement suggests, the anti-virus and anti-intrusion software should be kept up to date on a regular basis to prevent any known virus from doing any damage or stealing from the system.

6. Develop and maintain secure systems and applications

   Similar to keeping up to date with anti-virus software, it is important to install any patches and security upgrades as they become available. These would patch any holes within the software systems before they can be exploited. Of course the patches should be first deployed on a testing or staging environment to make sure that the applications would run correctly after the patch is applied.

7. Restrict to cardholder data by business need to know

Not everyone in your organization would need to know about user credit card information or user information. Only those individuals should be granted access to this data is who actually use or access this on a regular basis for defined business processes, such as running reports, customer service etc.

8. Assign a unique ID to each person with computer access

   While this in itself does not contribute to protecting the data, but assigning a unique id to each person would reveal the intruder or the account used for stealing or probing for data. This also means that one account shouldn't be used by multiple individuals. And the necessary precautions must be taken so that a user account is not compromised, such as enforcing the requirement for strong passwords, not sharing passwords, not writing passwords. The passwords must be encrypted during transmission and any remote access should only be allowed via unique tokens or unique certificates.

9. Restrict physical access to Cardholder data

   It is also important to restrict physical access to the cardholder data. Even if the significant data elements are securely encrypted, the removed data can be used for other malicious purposes. Furthermore any loss of data will negatively affect customer goodwill. This applies not only for servers but hard copies of data and backups.

10. <u>Track and monitor access to network resources and Cardholder data</u>

    It is important to keep track of the activities that a person or process performs. It creates an audit trail which can be referred to in case of a breach. Furthermore the audit trail can be monitored by automated tool which would warn the network administrators of attempts at unauthorized access and to detect patterns which would indicate a possible attempt at bypassing security.

11. <u>Regularly test security systems and processes</u>

    Testing the security of your system can reveal potential holes that might be exploited or reveal gaps in your logging capabilities to trace an intruder. By regularly testing your system using the latest exploits, you can ensure that your system is safe against newer types of attacks.

12. <u>Maintain a policy that addresses information security</u>

    There can be a lot of false positives in your attempt to secure your system. This might be due to lack of user knowledge and privileges about the system. A standard policy helps to educate users to reduce some of these false positives. Policies would also provide guidelines about non-computer aspects regarding security e.g. don't write passwords, don't share accounts, usage of flash drives, laptop protection etc. These preventative measures can secure your system against soft/social hacks.

More information going into the details on how to achieve all of the above requirements can be found at http://www.pcicomplianceguide.com/

5.10.4 Evaluation Processes

There are three different processes that need to be completed to demonstrate that the company is PCI Compliant. The process required for your company depends on the level that you would under. These are elaborated in the next section. The three processes are:

PCI Self Assessment Questionnaire: This questionnaire consists of 75 yes/no questions that must be filled in by the department that is responsible for processing the card information.

PCI Security Audit: This is done by a Qualified Security Assessors (QSA). This is an annual review to validate compliance in all applications and systems which store card information. It involves reviewing policy documents, sampling data to ensure it meets the PCI requirements and possibly taking interviews of personnel involved in the process.

PCI Security Scan: This entails a quarterly scan performed by an Approved Scanning Vendor (ASV). The ASV attempts to use the common and latest exploits against the payment system. This scan, which is conducted over the internet, is non-destructive to the system being tested. The result of the scan is a detailed report which identifies holes and vulnerabilities in the system (if any) and recommendations on how to fix the issues. If one of the vulnerabilities can allow a hacker to gain access to the card data or affect the system, the company is considered to have failed the scan.

5.10.5 Validation

After your organization has fulfilled the necessary requirements for PCI Validations, there must be regular checks performed on the system to find if the company is meeting the validation requirements. These requirements differ depending on the volume of transactions performed by the company each year. The current levels and their validation requirements are as follows:

Level 1- *Visa U.S.A. and MasterCard World Wide transactions totaling 6 million and up, per year, and any merchants who experienced a data breach.*

> These companies require an annual onsite review by merchant's internal auditor or a Qualified Security Assessor (QSA) or Internal Audit if signed by Officer of the company, and a quarterly network security scan with an Approved Scanning Vendor (ASV).

Level 2- *Visa and MasterCard transactions totaling 1 million to 6 million per year.*

> These require the completion of PCI DSS Self Assessment Questionnaire annually, and quarterly network security scan with an approved ASV.

eCommerce Best Practices

Level 3- *Visa and MasterCard e-commerce transactions totaling 20,000 to 1 million per year.*

> These require the completion of PCI DSS Self Assessment Questionnaire annually, and quarterly network security scan with an approved ASV.

Level 4- *Visa and MasterCard eCommerce transactions totaling up to 20,000 per year.*

> These require the completion of PCI DSS Self Assessment Questionnaire annually, and quarterly network security scan with an approved ASV. If a breach has been reported, or found, Visa can move the Level 4 merchant to a Level 1. If so, the Level 4 merchant must abide by the Level 1 validation requirements.

The full detail on the level of information that needs to be made available for the Approved Scanning Vendor and the Qualified Security Assessor is beyond the scope of this chapter. PCI Compliance has become a must. Failure to comply can result in surcharges, fines and increased liability if there is a data breach. It is necessary not only for security but also for your company's reputation.

5.11 SaaS vs. Managed Service

5.11.1   Introduction
5.11.2   SaaS (OnDemand)
5.11.3   Managed Service Providers and Perpetual Licensing
5.11.4   Budgeting for SaaS and MSP

5.11.1   Introduction

A company's IT budget as well as its business requirements are the two main factors in choosing between Software as a Service (SaaS), also known as OnDemand and ASP interchangeably, solutions versus Managed Service Provider solutions. These two models of service fulfill different levels of need in term of your commerce system.

The choice between SaaS and MSP depends on what type of business you run, how much infrastructure is needed to support your business processes, the skills you have in-house and the level your data is protected. SaaS is beneficial for companies who lack IT skill sets and for enterprises that do not require a highly customizable system. MSP, on the other hand, supports companies where there is a complex integration of systems to support the business.

Whereas SaaS is treated more as renting a system on a monthly basis, MSP suggests that the business owns the software. Deploying and maintaining the hardware and software for perpetual licenses can also create challenges for many companies. It requires a significant upfront capital investment, substantial in-house staff skills and a close working relationship with a variety of technology suppliers and service providers.

The evaluation and final decision to choose between OnDemand and MSP models usually depends on the architecture, infrastructure, manageability and total cost of owning the system. Companies need to measure against their business goals and whether an OnDemand system would be able to meet them. The price might be significantly less expensive than opting for an MSP model, but keep in mind that the functionality is also more limited. You have to project the ROI for both models and determine its worth.

5.11.2 SaaS (OnDemand)

The SaaS, OnDemand model of eCommerce systems is generally viewed as the low-cost alternative to actually owning the system with a perpetual license. However, the differences between the functionalities of SaaS and perpetual licensing systems are vast. Having a SaaS model forces you to work around the system; having a MSP model lets the system work around you.

SaaS is based on recurring monthly or annual subscription fees and typically follow a 'pay as you go' payment structure. The costs may increase as the usage of application increases. In some SaaS models, cost is dependent on number of users, transactions per day, SKUs etc.

A typical OnDemand model does not need any form of deployment, and can run over existing Internet access infrastructure. The model can be configured using APIs but multi-tenant application models cannot be completely customized. The OnDemand vendor assumes all training, infrastructure, support and security risks in exchange for reoccurring subscription fees. Since the model is designed to deliver business applications anywhere, anytime, this requires the platform vendor to employ a dedicated team that delivers the customers service levels.

Characteristics of an OnDemand Model:

- Periodic subscription based payment for the product—typically on a monthly or annual basis. Usually there is a one-time set up fee at the start of the project.

- Software is available over a shared network.

- Multi-tenant architecture—multiple customers run the same software but with different data.

- Product upgrades are enforced to all customers once the new release is ready to be launched.

- Template styled systems that severely limit customizing code.

- Relatively quick implementations—three to six months and can be done in remote locations, not on-site

- Hosting is done on the vendor's side, but could also be outsourced to a third party vendor.

5.11.3   Managed Service Providers and Perpetual Licensing

You would think that when a project takes two or three times to complete as originally estimated, the estimate would be deemed faulty. Wrong or right, that is not always the case. It is not always easy to determine what or who is at fault. Often the implementers, project managers and others who may or may not have had input into the estimate are faulted with their inability to produced the estimated results, overlooking scope creep, etc. If the estimator omitted large tasks from the estimate then that is where the blame will land. But more often than not, there are no such blatant omissions, yet the failure is in the estimation. After all, a good estimate will take into account scope-creep, the variance in productivity of different implementers, and in the case of eCommerce projects the likelihood that some requirements will change during implementation and shortly after what had been considered completion of the project.

Perpetual licensing in eCommerce refers to full ownership of your system, meaning the source codes are yours to change and customize. In order to receive that privilege you end up spending a considerable amount for the system. However, you will have greater control of the usability of your site and perpetual licenses do give you more flexibility to match your business requirements.

You can expect there to be significant upfront licensing costs and annual support costs. If you are expecting rapid growth from your eCommerce system, be sure it is scalable to account for that. Depending on the increase of number of users, prices may rise. The scope of your project will also affect the cost of ownership because there may be a need for additional hardware or IT support.

Traditional software tends to be highly customizable, with a typical package requiring hardware deployment, servers, backup and network provisioning, in order to accommodate the number of users on and off campus. Security architecture is also taxed to protect this valuable information from unauthorized access. The on-going maintenance and management of the application is provided by the customer's IT support teams installing the application on its network as well as providing for the logical and physical security to the applications, and the offering end-user training and support.

eCommerce Best Practices

Characteristics of a Perpetual License Model

- Software ownership. The customer makes a contractual commitment to owning the software, typically paying all - or a large part of the license fee upfront.

- Highly customizable to fit your business requirements, especially if those requirements are complex.

- Depending on the scope of the project, development and implementations are done by a consulting firm equipped to handle the customizations and integrations with different operation systems.

- Because of the opportunity to highly customize the system to fit your business goals, implementation can take anywhere from six to twelve months.

- Hosting is typically not included and will need to be handled client-side or by a third party vendor.

- Vendors maintain a release management framework, a substantial investment in installation and configuration tools, and substantial investment in QA to test every combination of infrastructure (hardware, database, middleware etc.) supported.

- Upgrades can be made when your business is ready, however they may be additional costs.

MSP solutions are needed for perpetual licensing because pf their complexity. It is individualized service, meaning not every client is built on the same hardware—it has its own separate hardware for your needs.

5.11.4 Budgeting for SaaS and Perpetual Licensing

The estimation technique expounded here is intended to be used prior to a detailed analysis and requirement definition phase where only high-level requirements have been identified. It is an accurate tool for measuring order of magnitude and has been employed in a number of projects with overall excellent results.

Budgeting is a key factor in deciding between SaaS or Perpetual Licensing. The normal IT budget can be broken up in three parts—the software that will do the processing, the hardware that the software needs to run on, and the staff that will support the software and hardware. A better description is given here:

- Software: The programs and data the organization uses for information processing and eCommerce platforms.

- Hardware: The computers, networking mechanisms, servers, and mobile devices that give support to the software and that allow people to access the information processing.

- IT Staff: Your employees, or people that your hire to operate the software and hardware, as well as maintain it usability.

The software is the most interconnected with your system's information management, which is why it is so important for you to select software that will suit your requirements. From that point, the hardware will fall in place in terms of knowing what is expected to support the software.

Your IT staff, or the consultants you outsource to, is also part of the budgeting process. You may discover that the staff you have in-house is not suitable to develop or implement the system you end up purchasing. If you do end up outsourcing to professional services, keep in mind the transition from using their services to keeping it in-house—knowledge transfer and training should be accounted for.

Perpetual Licensing Budgeting

Because of the scalable and customizable nature of owning software, it is often surprising for companies to see the amount of the budget going to professional services. However, this is important to note because it is in these services where the creation of your site's identity starts to take shape. This includes the design and development of the eCommerce site, as well as the integrations with other processing systems you may have.

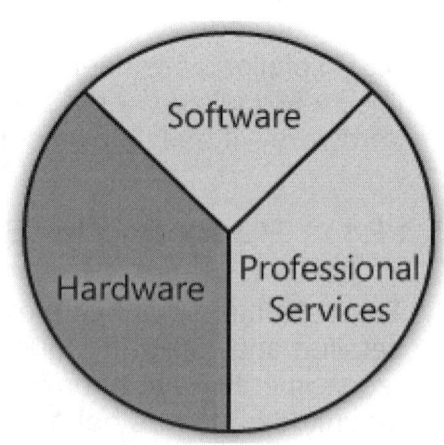

eCommerce Best Practices

The hardware budget is spent on servers to host data and applications, desktops and mobile computers. The software budget supports copies of licensed versions of business software, in addition to customized line-of-business software.

SaaS Budgeting

It should be mentioned that the overall SaaS budget is far less than normal perpetual licenses, sometimes reducing the cost over 80% depending on scope. Because SaaS solutions are already well established and the extent of customizations is fairly limited, the need for professional services is not as significant as with the perpetual licensing model. The professional services is there to basically get your site running and integrate it with other systems in place for your eCommerce site.

Since the hosted applications and servers are not located on the client's side, hardware support is also not a big factor in SaaS. Generally, it is the hosting vendor's responsibility to provide the hardware, software, and maintenance for the hosting environment.

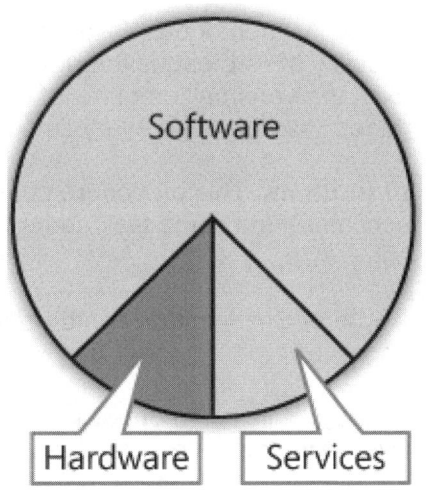

In the OnDemand model, the software budget is figured through the cost of subscription fees. This would be the majority of your budget for a SaaS model and proves to be a more cost effective system solution because presumably you would already have an IT staff in place, over the long-run, to maintain the site.

The case for SaaS or MSP models of service is dependent on your business requirements and the budget you are allocating for the eCommerce endeavor. There are benefits and misgivings to both models and it is ultimately your decision to evaluate the difference in ROI for the two.

GLOSSARY

Advergaming: Using games, specifically computer games, as a means to advertise or promote different products, organizations or viewpoints.

Adware: This is advertising supported software that automatically allows advertising material to be downloaded or displayed on a computer.

Affiliates: In eCommerce marketing, this is a website that when the product link is clicked, the website automatically links to another eCommerce website that sells the product from their site.

Aggregator: This is client software that is used to regularly check website updates, which then allows the content to be viewed together in one desktop application.

AJAX: Asynchronous JavaScript XML. Its purpose is to increase the interactivity, speed and usability of web pages through quicker responsiveness. Its programming allows for the exchange of small amounts of data with the server instead of having to wait for the user to reload the web page every time a change is made.

Algorithms: This procedure consists of using a finite set of well-defined instructions for accomplishing some task and will terminate the procedure once the final instruction completes.

Applets: Small components written in Java that can be "plugged" into a markup language page.

Analog: This is the opposite of digital and refers to information being presented continuously without the discrete on/off binary nature of digital.

Architecture of Participation: Describes the nature of systems that are designed for user contribution, such as open source and Wikipedia.

Atom: The name Atom applies to a pair of related standards. The Atom Syndication Format is an XML language used for web feeds, while the Atom Publishing Protocol (APP for short) is a simple HTTP-based protocol for creating and updating Web resources.

Avatar: Typically found in 3-D video games or online graphical chat rooms, avatars are virtual representations of a person or computer generated character.

Binary: A computer file containing any type of data, encoded to hold either one of two values: 0 or 1, On or Off, Yes or No, True or False. This byte format allows each character position to hold any one of 256 different binary codes. A series of bits make a byte.

eCommerce Best Practices

Bitmap: A data file or structure made up of generally rectangular grid of pixels, or points of color, on a computer monitor, paper, or other display device. Each pixel has a value representing black or white, a shade or gray, or a color.

Bluetooth: A telecommunications specification for a short-range radio frequency technology that helps coordinate mobile and fixed electronic devices. This enables mobile devices, i.e. cell phones, to wirelessly transfer information.

Blog: Personal home pages in diary format emphasizing community rather than commerce. Bloggers read others' blogs, link to them, reference them in their own writing, and post comments on each others' blogs. Blog is the shortened version of "web log".

Blogosphere: The collective term encompassing all blogs as a community or social network.

Blogroll: A list of links to other weblogs.

BOGO: Buy one get one free promotion.

BPEL: Business Process Execution Language. An orchestration language that uses serialized XML and which supports programming large concepts. This is the idea that BPEL language is necessary for high-level state interactions of a process, instead of language for single transactions.

BPM: Business Process Management. This is the plan and guide in how to manage business activity and how it should be performed.

BPMN: Business Process Modeling Notation. This is a standardized graphical notation used when creating a business process model.

Broadband: A signaling method which includes or handles a relatively wide range of frequencies which may be divided into channels or frequency bins.

C2C: Consumer-to-consumer. This is a form of eCommerce that uses the Internet to sell or re-sell items to other consumers.

C++: A general-purpose, high-level programming language with low-level facilities. It is a statically typed free-form multi-paradigm language supporting procedural programming, data abstraction, object-oriented programming, and generic programming.

CGI: Common Gateway Interface is a specification for transferring information between a web server and another program. A CGI program is any program designed to accept and return data that conforms to the CGI specification. Most commonly used languages are C, C++, Perl, or Java. However, CGI has some security weaknesses and does not scale as well as Java Servlets or similar multi-threaded direct-oriented architectures.

Channel Conflict: Bypassing Channel partners like distributors, retailers, sales reps and selling directly to the consumer through marketing or eCommerce.

Chatfly: A user who spends a lot of time in chat rooms and is willing to chat with anyone who will respond.

Citizen-journalism: The act of everyday people integrating themselves in the process of gathering, analyzing and reporting news and information.

Citizen-media: Any form of content produced by private citizens with a goal to inform and empower all members of society.

Cluetrain: This is the request for all corporations to speak on a human level and join in the global conversation that the Internet provides.

Collaboration: This is the act of a group of people, or groups of people, being able to work simultaneously and multi-task by way of a system like JSD. (See JSD).

Collective-intelligence: Similar to strength in numbers, this is the belief that large groups of people are smarter than a small batch of brilliant people.

Community / Participation: Social networks often spawn unique, specialized groupings of individuals with the same particular interests and/or backgrounds, which allows for a more participatory, open nature of the Web communication.

Content Management: Similar to Document Management but focused on text and media assets (GIF, JPEG, HTML, PDF, JavaScript, etc.) and the web upload process (view-in-context, staging, roll-back, etc.). This automates and controls the process of web content submission & publishing by authorized individuals within an organization without requiring technical web posting knowledge.

CORBA: Common Object Request Broker Architecture is a standard defined by the Object Management Group (OMG) to enable pieces of programs (objects or components) to communicate regardless of the programming language or operating system. Client applications can request services from the Object Request Broker (ORB) over the Internet Inter-ORB Protocol (IIOP).

Crowdsourcing: Obtaining labor, products, or content from a large group of customers, volunteers or amateurs who work for little or no pay.

CRM: Customer Relationship Management and eCRM (electronic CRM) is software, methodologies, and people intended to provide excellent, consistent experiences to an enterprise's customers. CRM spans many departments in an organization including customer support, sales, marketing, and accounting. Personalization and portal platforms are excellent eCRM solutions since they maintain consistent customer profiles and deliver targeted content to customers.

eCommerce Best Practices

CSS: Cascading Style Sheets, enhancements within HTML, allow web developers to define how standard elements such as headers and links should appear on all web pages using that style sheet.

Deeplinking: The act of a hyperlink on a web page that points to a specific page or image within another website, as opposed to that website's home page.

Digital certificate: A unique ID or "credit card" (typically following the X.509 standard) assigned to a user or application by a Certification Authority (CA) that includes the user's name, serial number, expiration date, encryption key(s), and/or similar information about the issuer.

Disintermediation: The removal of the middleman in the buying process, where customers and businesses freely interact without intermediary resellers or channel organizations.

Disruption: This is a spontaneous occurrence that causes disorder in information security. A disruptive technology is one that changes the dynamics of a market.

DNA: Windows Distributed interNet Applications Architecture, a marketing name for a collection of Microsoft technologies that enable the Windows platform and the Internet to work together. Some of the principal technologies comprising DNA include ActiveX, Dynamic HTML (DHTML) and COM.

DOM: Document Object Model, the specification for how objects in a web page (text, images, headers, links, etc.) are represented. The DOM defines what attributes are associated with each object, and how the objects and attributes can be manipulated. Dynamic HTML (DHTML) relies on the DOM to dynamically change the appearance of web pages after they have been downloaded to a user's browser.

DRM: Digital Rights Management is an umbrella term that refers to any of several technologies used by publishers or copyright owners to control access to and usage of digital data or hardware, and to restrictions associated with a specific instance of a digital work or device.

DSL: All types of Digital Subscriber Lines, the main categories being ADSL (asynchronous - different upstream & downstream speeds), SDSL (synchronous), and HDSL (High-data-rate). This copper wire connection uses sophisticated modulation schemes to connect the "last-mile" between telephone switching stations and homes/small offices. DSL can provide speeds from 64 Kbps to 6.1 Mbps or more over existing phone lines.

DTD: Document Type Definition is a type of file associated with XML and SGML documents that defines how the markup tags should be interpreted by the application processing and/or presenting the document. Schemas and namespaces are similar models for specific applications of XML.

Dynamic: A dynamic website is one that changes based on user actions or other factors. Dynamic sites often store content in a database rather than on individual static pages, allowing the site to be easily modified and used in a variety of ways.

EAI: Enterprise Application Integration is a business computing term for the plans, methods, and tools aimed at modernizing, consolidating, and coordinating the computer applications in an enterprise.

eBusiness: The process of using Web technology to help organizations streamline processes, improve productivity, and increase efficiencies. Enables organizations to easily communicate with partners, vendors and customers, connect back-end data systems and transact commerce in a secure manner.

Eclipse: A project supporting an open source integrated development environment (IDE). It provides a framework and a basic platform that uses plug-in software components provided by Eclipse members. Often used for Java development.

EDI: Electronic Data Interchange is the transfer of business data between organizations. EDI traditionally used the ANSI X.12 standard on expensive private networks but is increasingly occurring over the internet as B2B eCommerce, often using XML and encryption for secure data interchange.

EDMS: Electronic Document Management Systems manage the creation, modification, validation, and approval of documents (CAD files, SOPs, ECOs, forms, etc) in an electronic medium through electronic workflow. EDMS can improve document quality, reduce change turn-around time (i.e. time to market), ensure regulatory compliance for critical processes (e.g. pharmaceutical manufacturing), build a corporate Knowledgebase, improve employee productivity, and protect the security of documents.

EJB: Enterprise JavaBeans is a Java API developed by Sun Microsystems that defines component (reusable mini-program") architecture for multi-tier distributed systems. These platform-independent components are typically located on servers and often integrate a web front-end with back-end databases or legacy systems. EJB's allow developers to focus on the business architecture and provide a more scalable solution than CGI and/or Perl.

ERP: Enterprise Resource Planning systems are large business management software applications that integrate all facets of the business including planning, sales, purchasing, manufacturing, distribution, HR, finance, and marketing. ERP implementations are typically quite expensive and time consuming, due to their size and the implications in organizational change. However, as an integrated solution they can reduce data-reentry (errors & time), reduce time-to-market, improve operational efficiency, and provide an integrated supply chain.

Evangelists: Technical evangelists are those that promote new technologies through blogs, user demonstrations, articles, and other means.

eCommerce Best Practices

Exchange: An internet site that enables more efficient business transactions with financial savings by providing an open marketplace for buyers and sellers. For example, automotive, chemical, retail, or other industries could combine their collective buying clout to get supplies more cheaply and quickly while the manufacturer sells greater volume and realizes costs savings. This is different than Microsoft's Exchange product for email.

Executive Dashboard: An executive dashboard is a computer interface that displays the information corporate officers need to effectively run an enterprise. The term dashboard refers to a graphical user interface (GUI) that organizes and presents information in a format that is easy to read and interpret. (Also, referred to as an Executive Information System (EIS))

Extranet: An intranet that is partially accessible to authorized outsiders. Whereas an intranet resides behind a firewall and is accessible only to people who are members of the same company or organization, an extranet provides various levels of accessibility to outsiders (usually business partners or customers) based on privileges granted to a username/password.

Feeds: A web feed is a data format provided for serving frequently updated content. Content distributors syndicate a web feed, thereby allowing users to subscribe to it. Data feeds are also often used within eCommerce systems, for example a catalog update feed may update product data.

Firewall: A system designed to prevent unauthorized access to or from a private network. Firewalls can be implemented in hardware and/or software and are frequently used to prevent unauthorized internet users from accessing private networks connected to the internet, especially intranets. All messages entering or leaving the private network pass through the firewall, which examines each message and blocks those that do not meet the specified security criteria.

Flex: Adobe Flex is an umbrella term for a group of technologies initially released in March of 2004 by Macromedia to support the development and deployment of rich Internet applications based on their proprietary Macromedia Flash platform.

FOAF: Friend of a friend.

Folksonomy: A classification scheme in which Web users apply their own keywords to site content as a way of categorizing the data they find online.

Framework: Provides the structure on supporting the different aspects of a software project, albeit scripting language, code libraries, or support programs.

FTP: File Transfer Protocol is an application protocol that uses the internet's TCP/IP protocols to transfer files between hosts or download from a server. This protocol can be used from command line prompts, GUI FTP applications, or within a web browser.

eCommerce Best Practices

Globalization: The combination of the words global and localization. It is the intent of providing local service on a wider, global scale.

Heatmap: A measure of how much time a web user's eyes spend on certain sections of a web page.

Hibernate: Free Open Source software that provides an object-relational mapping, ORM, solution for Java language.

Hive: A tightly connected group of people who have similar interests, a similar culture, and talk to each other regularly.

HRMS: Human Resource Management Systems enable the HR department to create and manage forms and reports regarding personnel information independently. Typical functions include tracking employee benefits, time-off, job requisitions and applicant tracking, training, payroll, and employee and managerial self-service. HRMS may be integrated with a web portal.

HTML: HyperText Markup Language, the authoring language used to create documents on the World Wide Web. HTML evolved from SGML, although it is not a strict subset. HTML defines the structure and layout of a web document by using a variety of tags and attributes. The HTTP protocol is typically used to deliver an HTML file to a web browser.

HTTP: HyperText Transfer Protocol, the underlying protocol used by the World Wide Web. HTTP defines how messages are formatted and transmitted, and what action web servers and browsers should take in response to various commands.

HTTPS: HTTPS is a secure version of the HTTP protocol using SSL encryption and is often indicated by a key or lock in the bottom of web browsers.

ICE: Information and Content Exchange is an XML protocol that can be used to automate web content syndication (information sharing and reuse between web sites). ICE enables automated web asset management (supplying, exchanging, updating, controlling, etc) using a trust/security relationship based on the Open Profiling System (OPS). A different term "ice" (lower case) is an approach to controlling content placement on a web page that assumes left alignment and a fixed width in pixels. Unlike "ice", "Jell-O" pages are fixed width but centered and "liquid" pages resized to fit the width of the browser.

ICRM: Internet Customer Relationship Management improves the service provided to customers by personalizing the content on web sites based on the visitor's profile. Content can be targeted based on explicit information (e.g. filled out in a form) or implicit information (links they've previously clicked on) to improve the user experience (customer relationship). See also CRM.

IIS: Internet Information Server, Microsoft's web server that runs on Windows NT/2000 platforms (bundled with NT Server) but not on other operating systems. Active Server Pages (ASP), ActiveX controls, and FrontPage are well integrated with IIS.

eCommerce Best Practices

IM: Instant messaging is a method of typed-text, real-time communication between multiple parties.

IMAP: Internet Message Access Protocol is a way to access emails or messages that are kept on a mail server despite being at home, at the office, or traveling with a laptop. It allows you to access remote message stores as if they were local.

Infomediary: A composite of "information" and "intermediary" that describes web sites that aggregate information on specialized goods or services for a customer base for business transactions. An exchange is an example of an infomediary.

Infoware: Websites that use commoditized server software such as LAMP to enable data (e.g. book comments and ratings) to be shared via a website, and as a result create value (e.g. other people's opinions of a particular book that you want to buy).

J2EE: Java 2, Enterprise Edition enhances the Java 2 Standard Edition (J2SE) to provide a better thin client, multi-tiered architecture. Scalability, functionality, performance, and design capability are enhanced with: EJB, JDBC, JMS, JTA, JSP, JNDI, JavaMail.

Java: An object oriented programming language developed by JavaSoft (SUN) in 1995 that has had rapid adoption as the standard for eBusiness solution development. Although slightly similar to C++, it has many enhancements such as platform independence (write once, run anywhere).

JavaScript: A scripting language developed to enable web authors to design interactive sites. Although it shares some features and structures of the full Java language, it was developed independently by Netscape and is not compiled or object-oriented like Java. JavaScript can interact with HTML source code, enabling web authors to spice up their sites with dynamic content. JavaScript is an open language that doesn't require a license but sometimes operates differently in different web browsers (e.g. Internet Explorer, Firefox, and Netscape)

JavaBean: Re-Usable Java components ("mini-programs") that can easily be configured & integrated into other applications. PME - Properties (e.g. "user name"), Methods (e.g. "Update database"), and Events (e.g. "Mouse click") of one bean can be connected to another bean.

JDBC: Java Database Connectivity, a Java API that enables Java programs to execute SQL (Structured Query Language) statements. Vendor-specific drivers allow interaction with almost any SQL-compliant database from Java.

JINI: A "spontaneous networking" architecture developed by SUN to simplify connection and sharing devices (including wireless) on a network. Using Java and RMI, JINI enables devices (independent of O/S) to connect to a network without special device drivers.

JIT: Just-In-Time compiler, a code generator that converts Java bytecode into machine language instructions. Some Java Virtual Machines (JVMs), including the JVM in the Netscape Navigator browser, include a JIT compiler in addition to a Java interpreter. Java programs compiled by a JIT generally run much faster than when the bytecode is executed by an interpreter.

JMS: Java Message Service is an API from Sun that supports communication known as messaging between computers in a network. This common interface to both standard messaging and special messaging services supports asynchronous queuing.

JNDI: Java Naming and Directory Interface is an API with naming and directory functionality to applications written using the Java programming language. It is defined to be independent of any specific directory service implementation (e.g. LDAP, X.500, and Novell). Directory services can include information such as network printers, computers, and users.

JSD: Java Shared Data. This allows people to write distributed collaborative applications in order for groups of users to work simultaneously on a shared task.

JSR 168: A Java Portlet Specification that acts as a programming model for portlet developers and defines a contract between the portlet container and the portlets.

Knowledgebase- A repository of documents and other information stored in a centralized database or document management system from which individuals can upload, search, revise, and act upon group information. This collaboration can enhance and archive corporate (or other organization) knowledge and will grow as it is utilized.

LAMP: Linux, Apache, MySQL, PHP. A combination of OpenSource tools used as a web platform. LAMJ replaces PHP with JBOSS.

LDAP: Lightweight Directory Access Protocol is a protocol for accessing information directories (information about network users, computers, printers, etc) through a hierarchical tree structure. This lightweight (can run on the internet, intranet, PCs, etc) protocol is a subset of the X.500 standard and is supported by Netscape, Microsoft, Novell and other organizations.

Life hacking: The technical hacks that programmers have set up for in order to make their lives easier. This usually means methods of organizing data, little utilities to synchronize files, one-off scripts that automate daily tasks, etc.

Linklogs: A collection of URLs (hyperlinks) that the maintainer considers interesting enough to collect. Like a weblog, entries are listed in reverse chronological order. Unlike a weblog, though, postings are limited to just one link per posting and a title. Optionally, some further description or comment may be given.

LOK: Lack of knowledge.

eCommerce Best Practices

Long snout: Opposite to the long-tail, this term refers to the beginning stages of various products' development and demand curves, where millions of emergent products and technologies need to create early buzz and a lightweight means of publishing information about them.

Long tail: The internet phenomenon where millions of low-volume products at the end of their demand curve created a wider retail market. This term refers to the way in which the long tail attached to your computer mouse has led to a huge global market for obscure books, music, etc.

Mashups: This refers to websites that take useful applications from various websites and combine them in ways not intended by the original application. For example, mashup could combine mapping services such as Google Maps with business search services.

mCommerce: eCommerce transacted on a mobile phone or other device.

MD5: A message digest algorithm that analyzes and computes a specific "hash" from a certain input, which tests the integrity/value of data.

Meme: Pronounced meem. This is the concept of human culture evolving due to "contagious" communications throughout generations. Information is passed through imitation and behavioral replication.

Meta-tags: Descriptive labels applied to media assets, pages, information objects and/or learning objects that describe the object so it can be managed more effectively. They are machine-readable.

Microformats: Markups that allow expressions of semantics in an HTML (or XHTML) web page. Programs can extract meaning from a standard web page that is marked up with microformats. Adding microformats to a standard HTML web page allows machines to process HTML text and to possibly load data into remote databases. This would allow programs such as web crawlers to find items such as contact information, events, and reviews on web pages.

Micropersuasion: Using blogs and other interactive channels of specific target audiences to help organizations and organizations address certain issues and concerns, as well as capitalizing on the information/dialogue presented in those blogs.

Millenials: Generation Y (born around early 1980's-2001). The rising generation after Generation X, in which they're the most numerous, affluent, and ethnically diverse generation in history. Teenage Millennials are mostly the children of Boomers, preteen Millennials mostly the children of Generation X.

mmorpg: Massive Multi-Player Organized Role-Playing Game. This is an online computer game in which large numbers of players interact with each other virtually. World of Warcraft is an example of a mmorpg.

MMS: Multimedia Messaging Service. This is an enhanced service that allows video clips, graphics and sound files to be transmitted on a mobile or portable device.

MRP: Manufacturing Resource Planning applications utilize bill of material (BoM) information, master schedule, inventory, cost, and other information to optimize the manufacturing process. MRP is often a part of ERP systems, and is sometimes a source for product catalog data.

Moblog: A combination of the words mobile and weblog. This is a mobile weblog where content posted on the Internet can be viewed from a mobile or portable device, such as a cellular phone or PDA.

Monetization: The process of deriving revenue from groups of customers that would not normally be thought to attain revenue for the business. Monetization is often the goal of advertisements placed on websites.

MSM: Mainstream Media. This is commonly referred to "mass media" which is media designed to reach a wide audience that spans various demographics like, newspapers, magazines, TV advertisements and programs, music, and computer games. Specific Internet mediums include websites, podcasts, blogs, etc.

Multichannels: Leveraging multiple points of contact. For example, multiple channels for a retailer to interact with customers could interact with customers could include website, email, wireless (WAP, SMS/MMS, Call Center/CSR, printed catalog, kiosk, Point-of-Sale (POS), direct mailings, mass media, etc.

MVC: Model-View-Controller. This a design pattern that allows the separation of data and user interface through the use of the Controller.

NAS (Network Attached Storage): Data storage technology that allows heterogeneous network clients to access centralized data.

Netizen: A combination of the words Internet and citizen. This is a person who actively interacts with online communities.

Network: An interconnected system of computers and peripherals that communicates with each other.

OC-x: The Optical Carrier level indicates the signal rate (bandwidth) available on fiber optic backbones. The number "x" is a multiple of the base rate (OC-1) of 51.84 Mbps, so OC-3 is triple or 155 Mbps, OC-12 is 622 Mbps, etc. T1 (US standard: 1.544 Mbps also known as DS1) and E1 (European standard: 2.048 Mbps) may be twisted pair, coax, or fiber.

ODBC: Open DataBase Connectivity is a standard interface to databases developed by Microsoft. ODBC drivers are a middle layer between the application and the DBMS that translates the application's standard SQL queries into commands specific to that DBMS. It is

eCommerce Best Practices

also possible to use a JDBC-ODBC bridge to have Java applications communicate with ODBC-only databases (although most databases now have direct JDBC drivers).

OnDemand: Computer resources provided upon request and/or pay-per-use basis. Customers are only billed for using the resources, as opposed to having to own them. For example, a company can pay to lease software instead of buying a perpetual license. See also SaaS.

Open Lazslo: An open source platform for the development and delivery of web applications with a usable human interface (sometimes called Rich Internet Applications) on the World Wide Web. It is released under the Open Source Initiative-certified Common Public License.

Open source: Practices that advance access to the end product's source materials (i.e. source code) during production and development. For example, Open Source products include Linux, Apache, Eclipse, MySQL, PHP, and JBOSS.

OPML: Outline Processor Markup Language. This is an XML file format for outliner applications; however, it also is used to exchange RSS feeds between RSS aggregators.

Opt-in: When an individual chooses to receive emails or newsletters from organizations or corporations. Also considered "giving permission", Opt-out is the decision to not receive emails.

Pagerank: The key component of Google's web search metrics, which ranks the relative value of a page. The higher its value, the more visible the page in Google's search results.

PAN (Personal Area Network): A computer network that allows a person to connect to the Internet or a higher level network through computer devices, telephones, or other handheld devices.

Payment gateway: An application (sometimes based on the Secure Electronic Transaction (SET) protocol) designed to securely support financial transactions from web-based merchants. Its functions include authorization, payment capture, authorization reversal, credit, and settlement transactions. Transaction processing is often integrated into fulfillment processes for product delivery.

PCI: Credit card security/privacy regulations.

PPC (Pay Per Click): Sponsored advertising/listing on a search engine, usually located about or outside the natural search listings on a page.

Peer-to-peer: A network that relies primarily on the computing power and bandwidth of the participants in the network rather than concentrating it in a relatively low number of servers.

Permalink: A URL that directs you to an entry in a blog, even after it has been placed in the archives.

Perpetual alpha: Software or systems which are always in a state where they have full implementation of all its functionality but only satisfy a majority of the software requirements. (Not 100%).

Perpetual beta: Software or systems which are always in a testing phase.

Personalization: Helps the web better cater to the needs of the particular visitor to a web site by delivering content tailored to their interest. Visitors can be profiled explicitly (form-based data entry) or implicitly (e.g. tracking the links followed) to enhance the Internet Customer Relationship Management (ICRM) process.

Phishing: An illegal activity using social engineering techniques. Phishers try to obtain sensitive information like passwords or credit card numbers by conning people into thinking they are a trustworthy person or from a trustworthy business.

Podcast: Published audio feeds that can be transferred to an iPod or other digital audio player.

POP3: Post Office Protocol. It allows emails to be stored on a remote server and be retrieved at a later time.

Portal: Public portals generally synonymous with gateway, a web site that proposes to be a major starting point or one that a user tend to visit frequently as an anchor site. There are general public portals (e.g. Yahoo, Excite, CNET, MSN) and specialized or niche portals (e.g. Garden.com, Fool.com). Enterprise portals exist within a company and provide corporate and end applications to employees. Portals can also be used as Extranets.

PostScript: A page description language that uses a full programming language, rather than a series of low-level escape sequences, to describe the image being printed.

Profiling: A technique of observing user behavior where sets of characteristics of a particular web visitor are gathered from explicit data entry or observed (implicit) behavior. The collected information (profile) is very helpful in providing customized environments (content targeted to that profile) to maximize the online experience.

Prosumer: A combination of the words producer (or professional) and the word consumer. The line between producer and consumer is increasingly becoming blurred, as consumers provide more relevant content and advice online.

Prototype: Provides the framework for functions in developing JavaScript applications.

RDBMS: Relational DataBase Management Systems store data in related tables and typically accessed through Structured Query Language (SQL). Relational databases require few assumptions about how data is related or extracted from the database so data can be viewed and extended in many different ways. Relational tables contain columns of attributes and rows of data records.

eCommerce Best Practices

RDF: Resource Description Framework is a family of World Wide Web Consortium (W3C) specifications originally designed as a metadata model using XML but which has come to be used as a general method of modeling knowledge, through a variety of syntax formats (XML and non-XML).

REST: Representational State Transfer. This is a more lightweight protocol for Web Services than SOAP.

Reviews: Typically, subjective analysis of a product or service written by users/consumers or professionals/critics.

RIA: Rich Internet Applications. Unlike traditional web applications where all activity must be passed through a client-server, a RIA bypasses this by using a client engine (a layer of code) and allows interactions to be made on the client's computer.

RMI: Remote Method Invocation. This is the Java version of Remote Procedure Call (RPC), whether on the same machine or not, that allows Java code to pass objects along with requests to other Java code.

RMON: (Remote Network Monitoring) provides standard information that a network administrator can use to monitor, analyze, and troubleshoot a group of distributed local area networks and interconnecting T-1/E-1 and T-2/E-3 lines from a central site. Specified as part of the Management Information Base (management information base) as an extension of the Simple Network Management Protocol.

RoR - Ruby on Rails. This is an open source web application framework written in Ruby that utilizes less code with little configuration.

RSS: Really Simple Syndication or Rich Site Summary is a set of web feed formats used to publish frequently updated web pages.

SaaS: Software as a Service. This is a model of software delivery where the software company provides care, technical operations/instructions, and support for the software. Customers do not own a perpetual license. See also OnDemand.

SAN: A storage area network (SAN) is a high-speed special-purpose network (or subnetwork) that interconnects different kinds of data storage devices (e.g. disk drives and tape drives) with associated data servers. SANs can incorporate subnetworks with network-attached storage (NAS) systems.

Scenarios: As applied to personalized eCommerce, are customized sequences of targeted interactions that create persistent and personalized business relationships to drive enduring customer loyalty. Scenarios consist of goals, interactions, and metrics and extend beyond single web sessions to include the entire customer lifecycle and multiple channels (web, email, WAP, etc).

Scrum: A project management style intended for software development projects that can wrap extreme programming or other development methodologies.

SEO: Search Engine Optimization, focuses on improving the number and quality of visitors to a web site by making a site more machine-readable, and utilizing keywords in titles and copy. Used effectively, SEO results in higher visibility of a web site in the standard, "organic" results section of search engines.

SEM: Search Engine Marketing uses methods aimed at maximizing the visibility of websites in search engine results pages, utilizing both paid search inclusion and standard SEO techniques.

Semantic Web: An extension of the current Web that provides an easier way to find, share, reuse and combine information more easily. It is based on machine-readable information and builds on XML technology's capability to define customized tagging schemes and RDF's (Resource Description Framework) flexible approach to representing data. The Semantic Web provides common formats for interchange of data (where on the Web there is only had interchange of documents).

Servlet: Java code that runs on a server in a request/response manner to deliver interactive HTML web pages to client browsers yet still interact with back-end systems through the rich Java programming language. Servlets are more efficient than CGI scripts since they can be persistent and multi-threaded (won't consume one O/S process for each web visitor).

Smashups: SOA+AJAX+Mashups= smashups.

SMTP: Simple Mail Transfer Protocol is the most commonly used internet protocol for sending e-mail messages between servers. Messages can be retrieved with an e-mail client (MS Outlook, Eudora, Netscape Communicator) typically using either POP (Post Office Protocol) or IMAP (Internet Message Access Protocol).

Sniffer: A program that monitors and analyzes network traffic, detecting problems or occurrences that may arise. Sometimes also used by hackers for unethical purposes.

Sneezers: Enthusiastic early adopters of a product, service or experience. When they virtually "sneeze" they spread viral marketing.

SNMP: Simple Network Management Protocol (SNMP) is the protocol governing network management and the monitoring of network devices and their functions.

SOA: Service-Oriented Architecture explores the use of loosely coupled software web services to support the requirements of the business processes and software users.

SOAP: Simple Object Access Protocol (SOAP) is a way for a program running in one kind of operating system (such as Windows 2000) to communicate with a program in the same or another kind of an operating system (such as Linux) by using the World Wide Web's Hypertext Transfer Protocol (HTTP)and its Extensible Markup Language (XML)

eCommerce Best Practices

Social Bookmarking: Users store lists of Internet resources, which are then accessible to the public or to a specific network, so that other people with similar interests can view the links by category, tags, or even randomly.

Social network: Web-sites that allow users to "link" or interact with other users or members of the site. As memberships increase, the number of links in the network increases.

Spring: A layered Java application framework that offers a simple approach to development by cutting out numerous property files and helper classes.

SSE: Simple Sharing Extensions. These are RSS extensions that define the minimum extensions needed to enable loosely-cooperating applications.

SSL: Secure Sockets Layer is a protocol developed by Netscape for transmitting private documents via the internet. SSL works by using public and private keys to encrypt data that's transferred over a TCP/IP connection. Both Netscape Navigator and Internet Explorer support SSL, and many web sites use the protocol to obtain confidential user information, such as credit card numbers.

Struts: An Open Source framework for developing Java EE applications, and encourages use of the MVC architecture.

Supply Chain: Integrated Supply Chain Management (SCM) improves the core business functions of procuring, producing, and delivering products and services by better managing material, information, and fund flows. By managing an integrated supply chain (e.g. ERP systems), organizations can realize improved operational efficiencies, reduced costs, and faster time-to-market.

Symbian OS: An operating system designed for handheld devices that provide user interface frameworks and reference implementation of common tools.

Synergy: The act of human strengths and computer strengths working together. Computers process data more absolutely than humans, without common sense or judgments. Therefore, a person's thoughts are then translated into efficient processing of large amounts of data into a computer.

Tag Cloud: A visual emphasis of content tags used on a website. These are often designed for key information to stand out and catch the eye.

Tagging: The process of entering a keyword or term that is associated with or assigned to a piece of information or product/service that is desired, which enables a specific keyword-based classification of the information sought out.

TCB: Task Control Block. In the operating system kernel, this is a data structure that contains the information needed to manage/handle particular processes.

eCommerce Best Practices

Thought Leader: A person who is acknowledged among his or her peers for innovative ideas and demonstrates the authority to promote those ideas.

Trackback: An automatic reciprocal link to a site, usually included in the comments section of an article or entry.

UDDI: Universal Description Discovery and Integration is an XML-based registry (like a phone book) for businesses to list themselves on the internet and to enable system interoperability. UDDI is used by the SOAP web services protocol to enable automated transactions between organizations without extensive system integration.

UDP: User Datagram Protocol. This is a core protocol that allows programs on networked computers to send short messages to one another.

UML: Unified Modeling Language is a standard notation for developing object oriented programs. UML combines the Grady Booch Method of describing objects and their relationships, Jame's Rumbaugh's Object Modeling Technique (OMT), and Ivar Jacobson's Use Case Methodology, under the sponsorship of Rational Software.

User friendly: The ease with which people can navigate through a site or achieve a particular goal through the use of a certain product or tool.

Viral: Content posted on the Internet which gains widespread popularity through Internet sharing.

Vlog: The combination of video, web, and log. This is a blog that includes video medium.

Vodcast: Video podcast. The online distribution/delivery of OnDemand video clips.

VPN: Virtual Private Networks use the public internet to connect remote users as if on the internal LAN. The security of Point-to-Point Tunneling Protocol (PPTP) can be used to create extranets and wide area intranets instead of leasing private lines.

WAP: The Wireless Application Protocol is a secure specification that allows users to access information instantly via handheld wireless devices such as mobile phones, pagers, and two-way radios. Most operating systems support WAP although the typical handheld devices are expected to be PalmOS, Windows CE, FLEXOS, OS/9, and JavaOS. The layers of the WAP standard are: Wireless Application Environment (WAE), Wireless Session Layer (WSL), Wireless Transport Layer Security (WTLS), and Wireless Transport Layer (WTL).

Web 2.0: The second phase in the evolution of the World Wide Web where developers use new technologies to create websites that feel more interactive for users and look and act like desktop programs.

Web feed / Web syndication: A web feed is used as a part of web syndication. Web syndication is basically making a portion of a website available to use on other websites.

eCommerce Best Practices

Web server: A computer that is responsible for accepting HTTP requests from clients, which are known as Web browsers, and serving them HTTP responses along with optional data contents, which usually are Web pages such as HTML documents and linked objects (images, etc.).

Web Services: A software system that can be accessed through the Internet and executed on a remote system that hosts the requested services.

Web standards: The formal requirements and other technical specifications that define and describe aspects of the World Wide Web.

webtop / web desktop: A network application system that acts as a virtual desktop for the web. It integrates web applications into a web based work space.

Widgets: The specific ways a user can interact with the operating system and application depending on how the graphical user interface (GUI) is designed/displayed. This can include pull-down menus, selection boxes, progress indicators, windows, toggle buttons, etc.

WIFI: The technology of wireless local area networks.

Wiki: Websites containing online publishing tools so that any site visitor can contribute or edit the content.

WLAN (Wireless Local Area Network): A type of network that allows two or more computers to link without the use of wires. Because it uses radio waves as a means of connecting devices, the devices can be moved around with ease within a limited service area.

WML: Wireless Markup Language (formerly Handheld Devices Markup Language - HDML) allows the text portions of web pages to be presented on cell phones, PDAs, and other wireless devices. WML is part of WAP and runs on top of the GSM, CDMA, and TDMA data link protocols.

WSDL: Web Services Description Language is an XML-based language to describe how different web sites can perform operations (web services) on other web sites. WSDL files can be used over the SOAP, HTTP GET/POST, and MIME transfer protocols. WSDL can be used to record a specific web service available in the UDDI (Universal Description, Discovery and Integration) database.

Word of mouse: Communication through computer-based means, such as e-mail, social networks, or newsgroups.

Workflow: The operational movement of documents, data, applications, tasks, etc.

XSL: Extensible Stylesheet Language is a specification for separating style from content when creating XML pages. The specifications work much like templates, allowing designers to apply single style documents to multiple pages. XSL includes two major innovations --

allowing developers to dictate the way web pages are printed and specifications allowing one to transfer XML documents across different applications.

XML: Extensible Markup Language that allows designers to create their own customized tags, enabling the definition, transmission, validation, and interpretation of data between applications and between organizations. Unlike HTML's focus on just the presentation of information, XML better manages the data and allows presentation in multiple displays (e.g. HTML on a web page or WML on a cell phone, or VoiceXML over a telephone).

XHTML: Extensible HyperText Markup Language is a markup language that has the same expressive possibilities as HTML, but a stricter, wordier syntax.

XFN: XHTML Friends Network is a way to specify a relationship link to another person in the network, i.e. friend, classmate, sister, etc.

Yellow fade: A technique where the most recent item to change on a web page is highlighted by a yellow box which fades to the background color of the page after a couple seconds.

INDEX

Address Verification System. See AVS
affiliates, 43, 44, 45, 46, 47, 266
AJAX, 70, 243
API, 276, 281, 333
Application Programming Interfaces. See API
Applications, 35, 174, 193, 194, 216, 226
architecture, 194, 201, 202, 203, 205
   three-tier, 202
   two-tier, 202
Auto-completion, 277
Automated Inventory Management System, 222
AVS, 78, 82, 83, 90, 91
B2B, 60, 123, 124, 125, 126, 127, 128, 129, 166, 171, 261
back end financial processing systems, 82
BI, 229, 232, 233
blogs, 35, 36, 37, 38, 39, 48, 250, 251, 252, 253, 254, 255
   bloggers, 251, 252, 253, 255
   blogosphere, 252
   community, 251
   content, 252
   language, 252
   monitoring, 254
   netiquette, 255
branding, 2, 30, 31, 37, 142, 264
   personality, 34
business intelligence. See BI
business performance management. See BPM
Card Verification Method, 91
Card verification value 2. See CVV2

Cardholder Information Security Program. See CISP
catalogs, 162, 187, 222, 280, 284, 310, 313
charities, 121
chat, 15, 67, 244, 245, 246, 247, 248, 262
   instant messaging, 244
   Live Chat, 244, 245, 246, 248
   logging, 247
   monitoring, 247
checkout, 16, 32, 67, 68, 70, 71, 72, 74, 75, 76, 137, 145, 146, 313
   icons, 75
CISP, 78
click stream behavior, 305
click-thru, 43, 57, 306
CMS, 262, 264, 293, 294, 295, 296, 297, 298, 299, 300, 301, 302
   tools, 295
   WYSIWYG editor, 294
communication protocol, 101
community, 42, 45, 47, 115, 130, 170, 171, 264, 265, 266, 305
Content Administration, 296
   CA, 296
   revision control, 296
   role based personlization, 296
content management, 113, 132, 177, 212, 262, 264, 268, 293, 295, 296, 297, 300, 302
Content Management System. See CMS
conversion funnel, 231, 317
Coremetrics, 318
CRM, 19, 119, 143, 144

cross-sell, 16, 21, 55, 60, 61, 63, 128, 142
CSR, 156, 244, 245, 246, 247, 248
customer loyalty, 43, 44, 140, 142, 143, 169, 170, 245, 261, 304
   rewards, 141
customer ratings, 115, 118
customer reviews, 115, 116, 117
CVV, 327
CVV2, 79, 82, 83
data analytics, 229, 230
data integration, 191
Data migration, 285, 286, 289
   cleansing, 287
   incomplete, 289
   planning, 290
   post-migration, 292
   Scope, 290
   security, 291
donation, 121
dynamic content, 9, 294, 296, 299, 300
eBay, 38, 211, 213
eCommerce system, 208
EDI, 123, 129, 222
EFT, 222
Electronic Data Interchange. See EDI
Electronic Funds Transfer. See EFT
email marketing, 8, 10, 11, 13, 49, 212, 304, 305, 306
   automated emails, 10
   segmentation, 12
enterprise resource planning. See ERP
Enterprise Service Bus. See ESB
ERP, 127, 174, 261, 281, See Enterprise Resource Planning
ESB, 191, 194, 198
estimation, 268, 320, 321, 322, 323, 324, 326, 332, 334, 335

Extranet, 166, 167, 168
faceted search, 241, 242
Firewall, 216
Flex, 25
forums, 36, 39, 169, 170, 171, 172
   archiving, 172
   database support, 171
   flood control, 172
   karma points, 171
   polls, 171
fraud, 60, 81, 83, 87, 88, 89, 91, 92, 93, 94, 95, 98, 99
   Fraud Subsystem, 92, 93
   manual review, 92
   Negative Database, 88
   Positive Database, 88
   validation, 92
fulfillment, 15, 96, 97, 98, 99, 100, 101, 102, 103, 105, 117, 124, 126, 132, 149, 150, 151, 152, 153, 161, 212, 268
funnel navigation, 58
gift cards, 60, 75, 77, 80, 83, 84, 89, 135, 136, 137, 138, 139, 305, 306
gift certificates. See gift cards
gift services, 119, 120, 121, 122
giveaways, 306
Google Checkout, 210, 211, 213
GUI, 202
guides, 187, 306
hackers, 299, 352
hard goods, 130, 131, 132, 134
Hardware, 223
holiday, 121, 303, 304, 305, 306, 307
   special offers, 304
hosting, 191, 225, 226, 227
HTML, 28, 294
HTTP, 192, 193, 227, 257
HTTPS, 227

implementation, 25, 101, 103, 105, 107, 116, 134, 172, 192, 194, 216, 217, 222, 263, 272, 275, 320, 321, 322, 323, 324, 325, 334
   automation tools, 277
   build/deployment engine, 277
   deployment, 275
   design patterns, 275, 276
   reusing code, 276
   The Collections Framework, 276
infrastructure, 44, 126, 175, 192, 194, 222, 223, 224, 225, 226, 227, 230, 266, 297, 301
integration, 96, 132, 171, 174, 188, 190, 191, 194, 226, 228, 260, 261, 262, 263, 296, 310
inventory control, 108, 177
inventory management, 60, 102, 104, 105, 106, 108, 109, 129, 156, 161, 162, 163, 187, 279
inventory synchronization, 99
iTunes™, 133, 258
J2EE, 205, 265, 296, 297
key performance indicators. See KPIs
KPIs, 16, 232, 317, 318
labor management system. See LMS
LAN, 217, 223
LDAP, 264
legacy system, 285
LMS, 174
long-tail, 210
managed service, 214
Managed Service Provider. See MSP
merchandising, 60, 109, 111, 112, 114, 165, 303, 313
   dynamic landing pages, 113
Model and View Components. See MVC
MP3, 132, 133, 258

MP4, 258
MSP, 332, 333, 335
Multi-Channel, 151, 310, 311, 312, 313
   merchants, 306
multilingual, 296
MVC, 201, 203, 204, 205
NACHA, 85
National Automated Clearing House Association. See NACHA
network, 38, 42, 81, 109, 149, 150, 175, 191, 215, 216, 217, 223, 227, 244, 281, 282, 291, 299
Network Intrusion Detection Systems. See NIDS
newsletters, 12, 42, 45, 47, 54, 55, 56, 57, 58, 252
NIDS, 217
N-Tier architecture, 203
Offline, 30, 33
OLAP, 233
OLTP, 222
Omniture, 318
OnDemand. See SAAS
Online Analytical Process. See OLAP
online knowledgebase, 156, 157, 158, 159, 160
online manuals, 176, 177
Online Transaction Processing. See OLTP
operating system, 226
Operations
   logistics software, 282
   maintenance, 281
order management system, 96, 97, 149, 150, 152, 153
outsourcing, 105, 213, 214, 272
   offshore, 214
payment processing, 77, 78, 82, 104
   chargeback, 79

credit card, 81
forms, 80
Gateway, 78, 82, 226
Paypal, 85, 226
PCI Compliance, 326, 331
PCI DSS, 327
PCI Validations, 330
Perpetual licensing, 334
Personalization, 8, 10, 184, 234, 235, 238, 239, 261, 263, 266, 297, 299, 300
   click-stream data, 236
   explicit profile, 236
   repeat buyers, 234
   rule-based, 237
Platforms, 206, 226
podcast, 257, 258
podcasts, 39, 43, 47, 48
point-of-sale, 281
Portals, 184, 260, 261, 262, 263, 264, 265, 266
   B2B, 261
   B2C, 262
   B2E, 262
   benefits, 262
   collaboration, 262
   community, 264, 265, 266
   design and layout, 265
   framework, 264
   membership, 265
portlets, 260, 265, 266
product vendors, 162
promotions, 10, 16, 17, 19, 20, 21, 22, 23, 36, 55, 110, 142, 146, 170, 238, 239, 280, 284, 304, 305, 306, 307, 312, 313, 314
   internet kiosk, 19
Qualified Security Assessors, 330
Quality Assurance, 307
RDF, 159
Really Simple Syndication. See RSS
Representational State Transfer. See REST
Resource Description Framework. See RDF
REST, 192
returns, 102, 103, 107, 119, 151, 161, 162, 163, 164, 165, 210, 279, 280, 282, 306, 325
   cost, 163
   cross-channel, 164
   reasons, 163
RFI, 269, 274
RFP, 152, 153, 269, 270, 271, 272, 273, 274
   flaws, 271
   sections, 272
RIA, 24, 25, 26, 27, 28, 29, 35, 131
   AJAX, 25
ROI, 16, 17, 67, 268, 316, 318, 319
ROM estimate, 320, 325, 326, 332
RSS, 12, 21, 39, 43, 47, 48, 172, 256, 257, 258, 260, 261, 262, 263, 264, 266, 295, 296
   Atom 1.0, 257
   library, 257
   software, 256
RSS feeds, 21, 256, 295
SAAS, 175, 332, 333
scenarios, 2, 15, 16, 17, 18, 32, 77, 81, 82, 83, 84, 85, 94, 95, 102, 151, 224, 234, 237, 239, 297
schema design, 323
scope-creep, 321, 334
Search Engine Optimization. See SEO
search engines, 6, 226
search spiders, 5
searchandising, 113
security, 74, 78, 126, 127, 175, 187, 188, 192, 193, 194, 201, 215, 216,

217, 219, 222, 257, 263, 281, 297, 298
  access control, 171, 298
  database theft, 218
  intrusion detection, 217
SEM, 304
semantic web, 159, 160
SEO, 3, 4, 6, 7
  cloaking, 6
  Hidden keywords, 6
  Link Farms, 6
Service Level Agreement. *See* SLA
Service Level Requirements. *See* SLR
Service Oriented Architecture, 195, *See* SOA
service vendors, 207, 210, 212, 213, 214
shipping costs, 66, 71, 105, 120, 304
shopping cart, 10, 66, 67, 70, 71, 75, 145, 146, 186, 237, 239, 317, 323
Simple Object Access Protocol. See SOAP
SLA, 207
SLR, 207
small businesses, 210
SOA, 192, 193, 194, 195, 196, 197, 198, 199, 200, 260
SOAP, 191, 192, 193, 199
social networks, 35, 36, 38, 39, 40, 41
soft goods, 131, 132, 133, 134
  e-coupons, 132
Software as a Service. *See* SAAS
SPAM, 13
Stored Value Systems. *See* SVS

SVS, 83, 85
sweepstakes, 50, 51, 52, 53
targeted campaigns, 306
TMS, 174
transportation management system. *See* TMS
UDDI, 191, 193
UI, 185, 243, 298
Uniform Commercial Code, 85
Universal Description Discovery Integration. See UDDI
up-sell, 13, 63
user interface, 202
UX, 185
velocity checks, 90
warehouse control system. *See* WCS
Warehouse Management Systems. *See* WMS
warranty, 174, 175
  extended warranties, 174
WCS, 174
Web Analytics, 230
Web Services, 100, 191, 192, 196, 199
Web Services Description Language. See WSDL
Web Trends, 318
wikis, 38, 160
wish lists, 11, 67, 71, 146, 147, 148
WMS, 101, 174
word-of-mouse, 147
WSDL, 191, 192, 193
XML, 100, 126, 159, 191, 192, 193, 257, 296
XML Schema, 159

NOTES

NOTES